Metallic Multilayers and their Applications

Theory, Experiments, and Applications related to Thin Metallic Multilayers

HANDBOOK OF METAL PHYSICS

SERIES EDITOR

Prasanta Misra

*Department of Physics, University of Houston,
Houston, TX, USA*

Metallic Multilayers and their Applications

Theory, Experiments, and Applications related to Thin Metallic Multilayers

GAYANATH W. FERNANDO

University of Connecticut, Department of Physics,
Storrs, CT 06269, USA

ELSEVIER

AMSTERDAM · BOSTON · HEIDELBERG · LONDON · NEW YORK · OXFORD
PARIS · SAN DIEGO · SAN FRANCISCO · SINGAPORE · SYDNEY · TOKYO

phys
0 16301626

Elsevier
Radarweg 29, PO Box 211, 1000 AE Amsterdam, The Netherlands
Linacre House, Jordan Hill, Oxford OX2 8DP, UK

First edition 2008

Library of Congress Cataloging-in-Publication Data

A catalog record for this book is available from the Library of Congress

British Library Cataloguing in Publication Data

A catalogue record for this book is available from the British Library

ISBN: 978-0-444-51703-6
ISSN: 1570-002X

For information on all Elsevier publications
visit our web site at books.elsevier.com

Printed and bound in The Netherlands

07 08 09 10 11 10 9 8 7 6 5 4 3 2 1

The Book Series 'Handbook of Metal Physics' is dedicated to my wife

Swayamprava

and to our children

Debasis, Mimi and Sandeep

I dedicate this volume to my wife Sandhya and sons, Viyath and Surath

Gayanath W. Fernando

.

Preface

Metal Physics is an interdisciplinary area covering Physics, Chemistry, Materials Science and Engineering. Due to the variety of exciting topics and the wide range of technological applications, this field is growing very rapidly. It encompasses a variety of fundamental properties of metals such as Electronic Structure, Magnetism, Superconductivity, as well as the properties of Semimetals, Defects and Alloys, and Surface Physics of Metals. Metal Physics also includes the properties of exotic materials such as High-T_c Superconductors, Heavy-Fermion Systems, Quasicrystals, Metallic Nanoparticles, Metallic Multilayers, Metallic Wires/Chains of Metals, Novel Doped Semimetals, Photonic Crystals, Low-Dimensional Metals and Mesoscopic Systems. This is by no means an exhaustive list and more books in other areas will be published. I have taken a broader view and other topics, which are widely used to study the various properties of metals, will be included in the Book Series. During the past 25 years, there has been extensive theoretical and experimental research in each of the areas mentioned above. Each volume of this Book Series, which is self-contained and independent of the other volumes, is an attempt to highlight the significant work in that field. Therefore the order in which the different volumes will be published has no significance and depends only on the timeline in which the manuscripts are received.

The Book Series *Handbook of Metal Physics* is designed to facilitate the research of Ph.D. students, faculty and other researchers in a specific area in Metal Physics. The books will be published by Elsevier in hard cover copy. Each book will be either written by one or two authors who are experts and active researchers in that specific area covered by the book or by multiple authors with a volume editor who will co-ordinate the progress of the book and edit it before submission for final editing. This choice has been made according to the complexity of the topic covered in a volume as well as the time that the experts in the respective fields were willing to commit. Each volume is essentially a summary as well as a critical review of the theoretical and experimental work in the topics covered by the book. There are extensive references after the end of each chapter to facilitate researchers in this rapidly growing interdisciplinary field. Since research in various subfields in Metal Physics is a rapidly growing area, it is planned that each book will be updated periodically to include the results of the latest research. Even though these books are primarily designed as reference books, some of these books can be used as advance graduate level text books.

The outstanding features of this Book Series are the extensive research references at the end of each chapter, comprehensive review of the significant theoretical work, a summary of all important experiments, illustrations wherever necessary, and discussion of possible technological applications. This would spare the active researcher in a field to do extensive search of the literature before she or he would start planning to work on a new research topic or in writing a research paper on a piece of work already completed. The availability of the Book Series in Hard Copy would make this job much simpler.

Since each volume will have an introductory chapter written either by the author(s) or the volume editor, it is not my intention to write an introduction for each topic (except for the book being written by me). In fact, they are much better experts than me to write such introductory remarks.

Finally, I invite all students, faculty and other researchers, who would be reading the book(s) to communicate to me their comments. I would particularly welcome suggestions for improvement as well as any errors in references and printing.

Acknowledgements

I am grateful to all the eminent scientists who have agreed to contribute to the Book Series. All of them are active researchers and obviously extremely busy in teaching, supervising graduate students, publishing research papers, writing grant proposals and serving on committees. It is indeed gratifying that they have accepted my request to be either an author or volume editor of a book in the Series. The success of this Series lies in their hands and I am confident that each one of them will do a great job. In fact, I have been greatly impressed by the quality of the book "Metallic Multilayers" written by Professor Gayanath Fernando of the University of Connecticut. He is one of the leading experts in the field of theoretical Condensed Matter Physics and has made significant contributions to the research in the area of Metallic Multilayers.

The idea of editing a Book Series on Metal Physics was conceived during a meeting with Dr. Charon Duermeijer, publisher of Elsevier (she was Physics Editor at that time). After several rounds of discussions (via e-mail), the Book Series took shape in another meeting where she met some of the prospective authors/volume editors. She has been a constant source of encouragement, inspiration and great support while I was identifying and contacting various experts in the different areas covered by this extensive field of Metal Physics. It is indeed not easy to persuade active researchers (scattered around the globe) to write or even edit an advance research level book. She had enough patience to wait for me to finalize a list of authors and volume editors. I am indeed grateful to her for her confidence in me.

I am also grateful to Drs. Anita Koch, Manager, Editorial Services, Books of Elsevier, who has helped me whenever I have requested her, i.e., in arranging to write new contracts, postponing submission deadlines, as well as making many helpful suggestions.

She has been very gracious and prompt in her replies to my numerous questions.

I have profited from conversations with my friends who have helped me in identifying potential authors as well as suitable topics in my endeavor to edit such an ambitious Book Series. I am particularly grateful to Professor Larry Pinsky (chair) and Professor Gemunu Gunaratne (Associate Chair) of the Department of Physics of University of Houston for their hospitality, encouragement and continuing help.

Finally, I express my gratitude to my wife and children who have loved me all these years even though I have spent most of my time in the physics department(s) learning physics, doing research, supervising graduate students, publishing research papers and writing grant proposals. There is no way I can compensate for the lost time except to dedicate this Book Series to them. I am thankful to my daughter-in-law Roopa who has tried her best to make me computer literate and in the process has helped me a lot in my present endeavor. My fondest dream is that when my grandchildren Annika and Millan attend college in 2021 and Kishen and Nirvaan in 2024, this Book Series would have grown both in quantity and quality (obviously with a new Series Editor in place) and at least one of them would be attracted to study the subject after reading a few of these books.

Prasanta Misra
Department of Physics, University of Houston,
Houston, TX, USA

Volume Preface

Completion of this book has consumed several eventful years of my life with endless revisions and breaks but has been a fascinating experience. I could only marvel at the tremendous rate of experimental progress that has been achieved in this blossoming area of physics related to metallic multilayers and spintronics. The GMR effect, which was discovered only during the late 1980s, has paved the way for extremely tiny devices such as magnetic sensors, which in turn have made it possible to design high density data storage units that can be used, for example, in computers and iPods. It has taken only a decade or so for the GMR and related effects to change the way we live in this twenty first century. Its global impact has been remarkable which clearly points to the need for supporting novel and sometimes risky ideas in basic research. Unfortunately, there is an alarming trend, at least in the United States, which seems to be going against this obvious need for support. Throughout history, there are ample examples of serendipitous, scientific discoveries that have led the way towards providing significant benefits for the mankind.

This volume on metallic multilayers was written with undergraduate as well as graduate students and other researchers, who are new to this field, in mind. It does assume a basic knowledge of solid state physics, though certain chapters can be read only with some limited exposure to quantum mechanics. It certainly does not cover every aspect of metallic multilayers; I have chosen to focus on topics that could be of general interest, but there is naturally a personal bias in the selection of the topics and chapters. I would greatly appreciate any comments from the readers with regard to possible omissions, errors or anything else.

Many colleagues and friends have helped me in numerous ways to complete this volume and it is almost impossible to list everyone. However, I should mention Prof. Prasanta Misra, the editor of the book series, and Dr. Anita Koch at Elsevier, who have always been quite supportive and resourceful. My sincere thanks go to Prof. Gemunu Gunaratne at the University of Houston for helping us to initiate this work. At the University of Connecticut, several colleagues of mine, including Profs. Boris Sinkovic and Joseph Budnick, were kind enough to read certain chapters and provide valuable input. I have benefited greatly from my interactions with Drs. R.E. Watson, J.W. Davenport, Profs. M. Weinert and B.R. Cooper and this volume contains material from several projects in which we had a common shared interest. In addition, I am indebted to Dr. Mark Stiles for providing numerous relevant material which has made a real difference in this volume. I must also thank Dr. S.S.P. Parkin and Profs. Z.Q. Qiu and Y.Z. Wu for their help related to important topics covered here and Prof. W.C. Stwalley for constant encouragement during this project.

In addition, Kalum Palandage's assistance in getting numerous figures drawn is highly appreciated.

Last but not least, without my family's consistent help and support, a project of this nature would not have taken off the ground.

Gayanath W. Fernando
Storrs, CT, USA
December 2007

Contents

Chapter 1

GMR in Metallic Multilayers – A Simple Picture

1.1. Introduction

1.1.1. Nobel prize

The 2007 Nobel prize in physics was awarded to Albert Fert and Peter Grünberg for the experimental discovery of the GMR (giant magnetoresistance) effect, which is one of the main themes of this volume on metallic multilayers. The Royal Swedish Academy of Sciences, in making the announcement, stated that the Nobel prize had been awarded for the technology that is used to read data on hard disks. To quote the Academy, "it is thanks to this technology that it has been possible to miniaturize hard disks so radically in recent years. Sensitive read-out heads are needed to be able to read data from the compact hard disks used in laptops and some music players, for instance."

The GMR effect was a topic of fundamental research during the late 1980s and early 1990s that caught the attention of many and became an area of intense applied research. Within a relatively short period of time of this discovery, applications began to appear in the form of improved memory devices and sensors. This exciting area of research was named "spintronics" with electron spin dependent transport in metallic multilayers playing a leading role. Unlike some significant discoveries which are far removed from everyday life, this discovery has changed the world we live in, having already made a difference in the lives of ordinary people, especially with regard to information technology. The tiny memory devices that can store huge amounts of information, such as iPods, were made possible due to this discovery and its byproducts.

The basic physics related to the GMR effect is tied to the fact that the electron spin takes two different values (say up and down) and, when traveling through magnetized materials, one type of spin might encounter a resistance that is different from that experienced by the other. This property is referred to as spin-dependent scattering and can be exploited if electrical resistance is monitored in magnetic materials. Thin metallic multilayers became the proving ground for the GMR effect since their magnetizations could be altered with some ease, especially in a sandwich structure. In the following sections, more details can be found about the GMR effect, its discovery and history, magnetic recording as well as other related phenomena.

1.1.2. Magnetoresistance

Transport properties of metals and semiconductors have been of great interest to mankind for ages. Maxwell's equations, which unified electricity and magnetism, provided a rigorous set of rules governing electric and magnetic fields in a given medium. The Lorentz force **F** acting on a particle with charge q provided a "classical" picture of the effects of a magnetic field **H** in a nonmagnetic, homogeneous medium as,

$$\mathbf{F} = q\left(\mathbf{E} + \frac{\mathbf{v} \times \mathbf{H}}{c}\right). \tag{1.1}$$

A beautiful demonstration of an effect that could be related to the Lorentz force occurred in the discovery of the Hall effect in 1879, well before the development of modern quantum theory. Imagine a simple metallic strip carrying a direct current density (say j_x) in a given direction (x). In the presence of an applied magnetic field (H) perpendicular to the film (along z), a transverse current density (j_y along y) develops in the film (due to the Lorentz force, described above, acting on the charge carriers). This, in turn, will lead to a charge build-up on the boundaries leading to an electric field (E_y) opposing the transverse current. This field increases until the net force in the y-direction vanishes, and this is identified as the Hall effect. The Hall coefficient R_H, which determines the sign of the carriers, is defined as

$$R_H = E_y/Hj_x. \tag{1.2}$$

The component $\rho_{yx} = E_y/j_x$ of the resistivity tensor ρ is the Hall resistance. The diagonal component $\rho_{xx} = E_x/j_x$ is the magnetoresistance. Some of the early attempts to understand this effect in real materials were based on free electron theories. The magnetoresistance effects, i.e., changes in electrical resistance due to an applied magnetic field, are second order in H. Depending on the direction of the current measured and the external magnetic field with respect to the geometry of the metallic strip, one can identify various longitudinal and transverse resistivities. We define magnetoresistance ratio as a fractional change,

$$\frac{\Delta\rho}{\rho_0} = \frac{\rho - \rho_0}{\rho_0}, \tag{1.3}$$

where ρ and ρ_0 are the resistivities (along a given direction) in the presence and the absence of a magnetic field respectively. Simply put, the dependence of the electrical resistance of a material on an applied magnetic field (usually perpendicular to the direction of the current) represents this effect.

Magnetoresistance is, by now, a well-known transport property associated with materials. In this chapter, we will be discussing mostly metallic multilayers that exhibit drastic changes in magnetoresistance; hence the name giant magnetoresistance (GMR). The dependence of the electrical resistance of a metal on an applied magnetic field (usually perpendicular to the direction of the current) represents this effect. There is a class of oxides, known as CMR materials, which also exhibit colossal changes in magnetoresistance; we will not be discussing such materials in detail in this volume.

Translational symmetry in a periodic solid can be used to reduce the study of an (almost) infinite crystal to that of a small unit cell containing a few atoms. This enormous reduction in size was crucial to the early development of the quantum theory of the solid state. Associated with it is an important theorem due to Bloch, who introduced the **k**-vectors (called Bloch vectors) to describe the variation of the quantum mechanical wave function (state) throughout the periodic solid, once it is known inside the unit cell. Hence

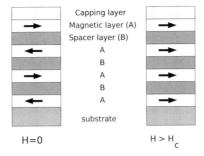

Figure 1.1. A typical multilayer sandwich structure and the changes in magnetization directions with an applied magnetic field **H**. The parallel magnetizations lead to a large magnetoresistance compared to that of the anti-parallel structure.

these **k**-vectors can be used to label the eigenstates of the Schrödinger equation. The electronic structure and transport properties in metals are understood to be closely connected to the so-called "Fermi surfaces" of metals. Such a (Fermi) surface can be thought of as a surface in k-space enclosing the occupied state **k**-vectors, acting as the boundary between occupied and unoccupied state **k**-vectors. The topology of the Fermi surfaces of metals is known to affect these transport coefficients due to the nature of the k-space orbits traversed by the electrons. In some metals, the magnetoresistance can grow large with increasing magnetic field, while in certain metallic multilayers, magnetoresistance was found to drop drastically compared to its zero field value.

A typical metallic multilayer unit consists of several layers of a metal A grown on several layers of another metal B (see Figure 1.1). This unit may be repeated so as to create a sandwich structure. In a magnetic multilayer system, one of the metals (say A) is necessarily magnetic while the other one (B) is nonmagnetic. The latter (B) is sometimes referred to as the spacer layer. Variations of the thickness of the spacer layer can give rise to dramatic oscillations in the magnetic coupling between the magnetic layers. These thin films can range in thickness from a few tenths of a nanometer to tens of nanometers. The magnetization directions of the ferromagnetic layers are coupled to each other through an exchange interaction. The sign of this coupling oscillates as the thickness of the spacer layer is varied, with the best multilayer samples showing up to thirty periods of oscillation.

1.1.3. Magnetic recording

An essential part of a modern computer is its hard disk drive (HDD) on which information is stored. Ever since its invention by IBM in the 1950s, the HDD has consistently outpaced Moore's law, which states that the storage capacity of computers will double every 18 months. Ever so often, there have been predictions about the maximum achievable recording density in recording devices, only to be surpassed by the next generation of such devices. Media for magnetic recording consists of thin, magnetically hard layers deposited on a substrate. In modern HDDs, the magnetic layer is a sheet of single domain grains. The grain structure inherits a certain randomness due to the manufacturing process; traditional magnetic recording deals with this randomness by averaging. A single unit of information, referred to as a "bit", will consist of a large number (hundreds) of grains whose average moment points along only two possible directions.

By 2003, the Giant Magnetoresistance effect had become, by far, the most practical application of metallic, magnetic multilayers [1]. The discovery of large changes in magnetoresistance and oscillatory behavior of exchange coupling in transition metal multilayers had paved the way for new classes of thin film materials, suitable for magnetic sensors as well as magnetic random access memory (MRAM) based devices. Magnetic recording, at a reasonably competitive level with other recording options, became a reality with the invention of the GMR devices (see Figures 1.2 and 1.3).

In a HDD based on magnetic recording, information is stored by magnetizing regions within a thin film. The transitions between such regions represent "bits" which are detected by a "read head". The number of magnetic bits per unit area, called the areal density, has increased dramatically during the past decade. This is mostly due to magnetic recording read heads based on the GMR effect, which were first introduced by IBM in late 1997 [2]. Such read heads became the industry standard in 1999 and could be found in virtually all hard disk drives, which were assembled during to period 1999–2005. Numerous technical hurdles have been overcome in reducing the bit size in the search for high recording density. These problems range from read issues such as the magnetic shielding and sensor sensitivity; write issues, such as magnetic field strength and speed; media (i.e., HDD) issues, such as noise and data stability as the bit size approaches the superparamagnetic (SP) limit, which is reached when the ambient thermal energy $(k_B T)$ becomes comparable to the magnetic energy KV, V being the grain volume and K the

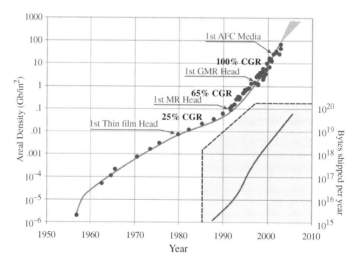

Figure 1.2.   Increases in areal density and shipped capacity of magnetic storage over time. Since the invention of magnetic recording in the 1950s, the areal density (bits stored per square inch) of disk drives has increased rapidly. Fueled by the development of the anisotropic magnetoresistance (AMR) and GMR based sensors, in recent years its compound growth rate (CGR) has exceeded 100%. Developments in the magnetic media have kept pace with those of the read sensor and write head. For example, antiferromagnetically coupled (AFC) recording media allow the writing of narrower tracks on media which would otherwise be susceptible to superparamagnetic effect. The inset plot shows the total capacity of hard drives shipped per year; in 2002 that shipped capacity was 10 EB worth of data. (Reproduced from Ref. [1] with permission from © 2003 IEEE.)

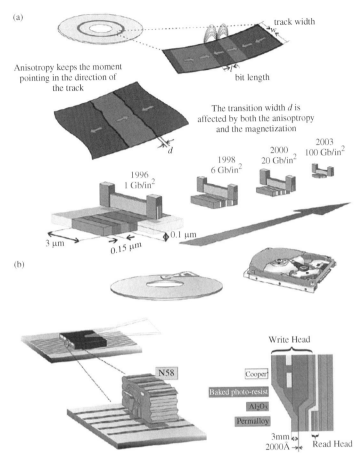

Figure 1.3. Fundamentals of magnetic recording: Schematic diagram of a hard disk drive. (a) Information is stored by magnetizing regions of a thin magnetic film on the surface of a disk. Bits are detected by sensing the magnetic fringing fields of the transitions between adjacent regions as the disk is rotated beneath a magnetic sensor. As the area of the magnetized region has decreased, the read element has to scale down in size accordingly. (b) The read element is incorporated into a merged read-write head, which is mounted on the rear edge of a ceramic slider flown above the surface of the rapidly spinning disk via a cantilevered suspension. A hard drive unit usually consists of a stack of several such head-disk assemblies plus all the motors and control electronics required for operation. For more details, see Ref. [5]. (Reproduced from Ref. [1] with permission from © 2003 IEEE.)

anisotropy constant. Another significant challenge, as the bit size gets smaller and smaller, is to extract a strong enough magnetic signal from a tiny, magnetized region [1].

Information stored using magnetization of thin film regions (on the surface of a disk) of nanometer length scale would naturally lead to high density storage. However, such a film has to satisfy several requirements. (a) The average magnetization has to point along the direction of the track in only two possible directions and the anisotropy should be strong enough to keep the moments in plane as described above (longitudinal recording); (b) A small region with a moment pointing along one direction will be identified as a bit;

(c) Adjacent bits or demagnetizing fields cannot affect the moment of a given bit; i.e., coercivity of a bit must be sufficient to overcome demagnetizing fields; (d) Bits are detected by sensing the magnetic fringing fields of the transition fields between the adjacent regions as the disk is rotated underneath a magnetic sensor; (e) The read element which is embedded into a read head has to scale (or match) with the size of a bit (otherwise, the stored information would be averaged out!).

It is not surprising that the highest GMR effects are seen in magnetic multilayer structures containing the thinnest possible magnetic and nonmagnetic layers [1]. For technological applications this is considered a useful property. Demagnetizing fields, which depend on the shape and size of a given magnetic material, tend to play the role of a spoiler in the context of applications. This is because such demagnetizing fields, associated with the ferromagnetic layers, will eventually make it necessary to have large external fields in order to change the magnetic state of a device. In sensors, this will lead to limitations of the smallest detectable field. For this reason, it is desirable to have smaller magnetic (grain) volumes. In memory applications, demagnetizing fields will eventually limit the memory density. It should be clear from the above discussion that controlling the demagnetizing fields and grain volume plays an important role when fabricating practical devices.

1.1.4. Perpendicular recording and future storage technology

At present, the areal densities achieved using longitudinal (i.e., parallel to track) recording range from 100–150 Gb/in^2. The superparamagnetic limit, mentioned above, is reached when the ambient thermal energy ($k_B T$) becomes comparable to the magnetic energy KV, V being the grain volume and K the anisotropy constant. This situation will clearly lead to instabilities in magnetizations and hence, data cannot be stored reliably. If the magnetic moments are aligned perpendicular to the surface of the media, there will be a clear advantage of capacity. Research efforts along these lines have been progressing over the past decade or so [3] and, with perpendicular recording media, the areal densities up to about 250 Gb/in^2 have been achieved. Again the SP effect is expected to limit the areal densities in perpendicular recording to about 0.5–1.0 Tb/in^2.

Perpendicular media, mentioned above, is merely another stop-over in the search for higher areal density. Two new approaches currently being considered are, heat-assisted magnetic recording (HAMR) and recording on bit patterned media (BPM). In BPM, the aim is to make a single magnetic grain the object of recording bits and both these approaches are targeting areal densities beyond 1 Tb/in^2 (see Figure 1.4). Currently, one bit in Seagate's 250 Gb/in^2 HDD contains about 65 grains [4]. More detailed information on magnetic recording can be found in Refs. [3–5].

Magnetic sensors based on spin-valve GMR sandwiches have resulted in gigantic increases in storage capacity of magnetic HDDs. The read signals from GMR devices is one to two orders of magnitude higher than those from the previous generation of read heads, based on the anisotropic magnetoresistance (AMR) effect [6]. The AMR effect (of a few percent) is quite small compared to GMR (usually 100% or more at room temperature as illustrated in Figure 1.5), and is dominated by bulk scattering, in contrast to interface scattering which is believed to be predominant in GMR. Magnetic Tunnel Junction (MTJ)

PARALLEL

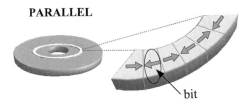

bit

PERPENDICULAR

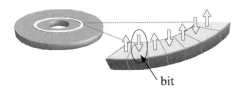

bit

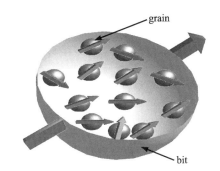

grain

bit

BPM

Single grain

Figure 1.4. A cartoon comparing parallel and perpendicular recording media as well as depicting bits and grains (not to scale). The magnetic moment of a given bit is obtained by averaging the moments of individual grains that constitute the bit. In BPM (bit patterned media), the aim is to make a single magnetic grain the object of recording bits.

devices have made it possible to develop advanced MRAM with densities close to that of dynamic RAM and read-write speeds comparable to static RAM.

These particular examples illustrate spin dependent transport properties in artificial structures (thin film devices) that can be controlled by an external magnetic field. They also show new physics at smaller length scales (i.e., at small film thicknesses) as well. This is also an example of concise experiments that led the way and needed theoretical understanding. At this stage, it is worthwhile and instructive to examine the seminal work that was progressing in the mid to late eighties and well into the nineties.

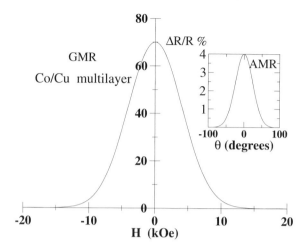

Figure 1.5.   Schematic comparison of GMR and AMR. The anisotropic magnetoresistance, which is at least an order of magnitude smaller than the giant magnetoresistance, is mainly due to bulk scattering of spin-polarized electrons from a ferromagnetic material. The angle between the current and the applied magnetic field is denoted by θ. The GMR effect is now understood to be closely tied to interface spin scattering. (Following Ref. [1] with permission from © 2003 IEEE.)

1.2.   GMR: A historical perspective

The early ideas related to spin-dependent scattering can be traced back to Mott in the 1930s [7], who recognized that in a ferromagnetic metal, electrons with different spins would experience different resistivities. Later, in the 1960s, Fert and Campbell [8], among others, recognized the importance of this distinction. In addition to Fert and Grünberg who were awarded the 2007 Nobel prize, many others have contributed to the rapid development of this technology; among them, the experimental group at IBM (Almaden) led by Stuart Parkin played a significant role in bringing the discovery to the next level.

In 1986, Peter Grünberg and his group [9] investigated the magnetic coupling of Fe layers separated by layers of Au and Cr using light scattering. Single crystal growth along the (100) direction was essential here. For Cr interlayers of an appropriate thickness, it was found that the Fe layers on either side of Cr, with Fe(120 Å)/Cr(10 Å), were aligned antiferromagnetically. This behavior was quite different from that due to Au, where the coupling strength decreased continuously (normal/the expected behavior) as the Au thickness was increased from 0 to 20 Å. In the same year, there were other similar reports based on rare-earth yttrium multilayers [10,11]. Several other papers followed the original work of Grünberg *et al.* around this time [12]. Notably, the French group led by Albert Fert [13] was working independently and found similar coupling. Magnetoresistance measurements on several metallic bilayer systems were reported later from 1988 to 1990 (by Baibich *et al.*, Grünberg *et al.* and Parkin *et al.*) [13–15].

The most decisive and significant changes, as large as 80% in magnetoresistance, were first reported by the French group (Fert and his collaborators) in 1988 for Fe(30 Å)/Cr(9 Å) multilayers deposited on a GaAs substrate. The metals used in the original experiment of Baibich *et al.* [13] were Fe and Cr (001) bilayers. These multilayers were prepared by MBE (molecular beam epitaxy) at low pressure ($\simeq 5 \times 10^{-11}$ Torr) and temperature. In the absence of an applied field, the magnetoresistance was found to increase with decreasing Cr layer thickness, attaining the observed maximum for Fe(30 Å)/(Cr 9 Å). At this thickness, the (adjacent) Fe layers separated by Cr were found to be antiferromagnetically coupled. By slowly applying an external magnetic field, one can try to obtain the field necessary to saturate the magnetization. These saturation fields provide a measure of the strength of the antiferromagnetic coupling (of Fe layers separated by Cr) at zero field, i.e., the exchange energy. The highest saturation field (about 20 kOe) was found for the bilayer thickness given above, i.e., at 30 Å / 9 Å Cr value. Increases in the Cr thickness while maintaining the Fe thickness at 30 Å yielded smaller saturation fields. The clear conclusion was that the maximum in the Magnetoresistance occurred when the adjacent Fe layers separated by Cr were (completely) antiferromagnetically coupled; i.e., when the exchange coupling of the adjacent layers showed oscillatory behavior.

There were many open questions that had to be answered at this stage. What is the scattering mechanism responsible here to change the resistance as observed? Clearly, there is a spin dependence in the transport of electrons. What other materials would show a similar effect? What determines the strength of the saturation fields? How important is the interface scattering? All these issues have been addressed to varying degrees of satisfaction. There were two observations that seemed imperative for large changes in magnetoresistance: antiferromagnetic coupling through the spacer layer and the ability to switch this coupling from AFM to FM with an applied magnetic field.

Early theoretical attempts to understand the oscillatory behavior followed free electron theories, but the oscillatory periods were usually longer than the nominal free electron periods in Ruderman–Kittel–Kasuya–Yosida (RKKY) coupling [16]. The spatial distribution of magnetization in the neighborhood of a localized (magnetic) moment can be shown to oscillate falling of as $1/r^3$ at large distances r. The oscillations show a wave-number $2k_F$ associated with the free electron Fermi wave vector and short oscillation periods of the order of a few Å. This observation resulted in early doubts about whether the oscillations in coupling strength are in anyway related to the RKKY theory [17]. The general understanding that the changes in the density of states due to multilayering were responsible for the oscillatory behavior was not always there. Some argued that the bulk impurity (defect) scattering was the primary cause of the GMR effect. The importance of the interface effects were pointed out decisively by Parkin *et al.* [15] in their beautiful work; high quality interfaces were clearly responsible for obtaining large increases in the magnetoresistance. Of course, when the thickness of the nonmagnetic layer approaches the bulk value, one would expect bulk-like scattering to play a central role, but this is not to be confused with the comparatively larger GMR effect due to interface scattering. Also, there were models that were based on single band and multiband ideas. Can Fermi surfaces derived from bulk metal calculations really explain the oscillatory coupling? Confined states due to the presence of magnetic/nonmagnetic interfaces as described above are now recognized as being responsible for large changes in magnetoresistance.

By 1993, the oscillatory exchange coupling had been measured in a large number of multilayer systems. Theoretical understanding was also evolving in parallel with the ability to grow high quality samples and make accurate measurements. Systematic sputtering based studies, with Co as the ferromagnet, had revealed periods of about 0.9 to 1.2 nm for V, Cu, Mo, Ru, Rh, Re, and Ir spacer layers and longer periods for Os (1.5 nm), and Cr (1.8 nm) (Refs. [18–24]). On the other hand, lattice matched epitaxial systems showed more complicated behavior including a short period (examples include Cu/Co, Ag/Fe, Cr/Fe). By 1993, there were several theoretical models that tried to explain the oscillatory coupling and most of them were pointing to the an important fact that the Fermi surface of the *spacer layers* determined the period of the oscillations.

However, in 1993 there was no consensus with regard to the electronic structure mechanism that was responsible for the interlayer exchange coupling. This was mostly due to two competing ways of understanding ferromagnetism in these transition metals. A localized description of ferromagnetism assumes that the magnetic moments, mostly due to d electrons, are local and hybridization of such d states with s and p electrons, which are itinerant, results in ferromagnetic (long range) order. The Anderson model was developed, at least in part, to address similar situations and provides a pertinent example. An alternate, band-theory based description treats all the valence electrons as delocalized and that there are spin-split d bands due to exchange interaction which give rise to (long range) ferromagnetism. The Fermi surfaces calculated from these two different descriptions will not necessarily be the same.

Short circuit effect was a popular way used to explain the preferential spin scattering. This is a rather simple minded way of beginning to explain the GMR mechanism. When the layer coupling is anti-parallel, a given electron of parallel or anti-parallel spin is likely to experience the same, average electrical resistance, while if the magnetic layer coupling is parallel, then a parallel spin feels a lower resistance (less scattering) compared to an anti-parallel spin. Hence, the short circuit effect occurs; i.e., current is carried mainly by parallel spins. However, this does not answer the question completely. Why do the parallel spins see a lower resistance? A detailed answer to the above question begins with an examination of the electronic states which participate in the transport process.

1.3.   Qualitative arguments

Such states are found in the vicinity of the Fermi level (or Fermi surface). In a typical $3d$ transition metal, d states dominate the density of states around the Fermi level, while there are s and p states in the same energy region that are spatially more extended. A simple, qualitative argument for differential scattering (seen for up and down spins) can be given as follows. The s or p electrons, which are closer to being free electrons than the d electrons, are thought to carry the current, while the role of d electrons is tied to scattering. The electrons that get scattered fall in to empty d states and in the ferromagnetic $3d$ metals, there are less d holes (empty states) in spin up (majority) states when compared with spin down states. For example, Ni and Co being strong ferromagnets in this sense, when compared to even Fe, act as better GMR materials, since they have less holes in the majority d states which prohibits scattering in to them as efficiently as in to minority d states. From the above discussion, one can see that subtle band structure and Fermi surface effects can be responsible for increases or decreases in the GMR effect. Band matching across an interface can also play an important role. Also, the fact that $4d$ or $5d$ metals do not seem to play a role in GMR points to the absence of magnetism as well as larger spin-orbit effects.

Fermi surface effects naturally remind one of the de Haas–van Alphen oscillations closely tied to measuring Fermi surfaces. In brief, one can use the oscillations in MR observed with respect to an applied magnetic field. These help in mapping out extremal areas associated with the Fermi surface along the direction perpendicular to the applied field. The oscillations are related to the closed orbits (in k-space) that an electron can traverse before getting scattered. For typical fields, these orbits, in real space, can span large regions and hence require long mean free paths. This implies that unless high quality, clean samples, devoid of impurities, are used, it will be difficult to make use of such orbits. Interlayer exchange coupling can also be qualitatively regarded as somewhat similar to this situation in that the electrons must complete a round trip in the spacer before getting scattered. However, this path is much shorter than the typical paths for the de Hass–van Alphen effect, which is consistent with the fact that the GMR effect is observed at temperatures higher than that of the room temperature. These ideas are related to the necessity of high quality samples in order to observe the GMR effect.

Another intensely debated topic, especially in the early days, was the location of the scattering centers; i.e., whether they were located at the interfaces or not. Now, it is generally believed that the interface effects (scattering etc.) are mainly responsible for GMR. During the early days, the role of the interlayer coupling was also not quite clear. Here Parkin (Ref. [18]) was able to point out that antiferromagnetic coupling across the non-magnetic material was a very general occurrence, not limited to Cr.

1.3.1. Quantum well (QW) approach

A simple quantum well approach can also yield some important insights into magnetic or nonmagnetic multilayer systems. This elementary approach, familiar to those with a basic background in Quantum Mechanics, can be applied with certain modifications. One essential aspect of the QW ideas is to find a single particle potential appropriate for various regions of space that are being probed. For transition metals, which are quite different from free electron like simple metals, this task may not appear that trivial at first sight. We also know that magnetic interactions are inherently two-body interactions. Fortunately, there is a very well tested method, based on the density functional theory (DFT), to obtain such potentials when dealing with ground state properties of transition metals. Although the reduction of the many-body problem to a single particle one is nontrivial, for ground state properties of metallic systems, this method has been highly successful. An insightful discussion of DFT is provided in a different chapter of this book.

1.3.2. Spin-polarized quantum well states

Spin-polarization or spin-splitting of energy band states gives rise to magnetic moments in an itinerant model of electronic structure. These moments arise due to the presence of more occupied spin-up states (say) compared to the spin-down states. In this section, the properties of a magnetic multilayer using a spin-split, free electron model will be discussed following Ref. [25]. In such a model, interfaces are simply potential steps. In later chapters, we will discuss more realistic, band structure based, models of the electronic structure. An interface between two different materials behaves like a potential step as far

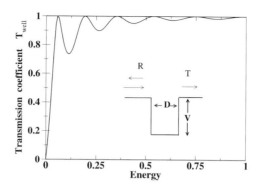

Figure 1.6.   Transmission coefficient as calculated for parameters ($V = 0.25$, $D = 20$ in atomic units) listed in text.

as the electrons are concerned; in general, electrons that strike it have their transmission probability reduced.

As discussed in elementary quantum mechanics texts, the probability of transmission (or reflection) can be conveniently expressed in terms of the transmission (or reflection) coefficients. For a free electron going down a simple potential step of height V (> 0), the transmission coefficient is,

$$T_{step} = \frac{q}{k}\left(\frac{2k}{k+q}\right)^2, \tag{1.4}$$

while the reflection coefficient is

$$R_{step} = \left(\frac{k-q}{k+q}\right)^2. \tag{1.5}$$

Here, the incident wave vector is $k = \sqrt{2mE/\hbar^2}$ and the transmitted wave vector is $q = \sqrt{2m(E+V)/\hbar^2}$.

The first factor accounts for the change in velocity on going over the step. If another step is introduced, the electron undergoes multiple reflections inside the quantum well that is formed. If the steps are separated by a distance D, the transmission probability (for $E > 0$) is (see Figure 1.6),

$$T_{well} = \left|\frac{4kq}{(k+q)^2 - e^{i2qD}(k-q)^2}\right|^2. \tag{1.6}$$

Note that there is perfect transmission (i.e., probability $= 1$) whenever $2qD = 2n\pi$, where n is an integer. This condition, at a fixed thickness, gives rise to a series of trans-mission resonances at the energies $E_n = [\hbar^2/2m](n\pi/D)^2 - V$, for integer n. At a fixed energy, there are resonances for $D = 2n\pi/(2q)$ with $q = \sqrt{2m(E+V)/\hbar^2}$; these reso-nances are separated by a fixed increment in thickness. At such transmission resonances, the probability of finding electrons in the quantum well increases, i.e., there will be a peak in the density of states in the quantum well at the energy of the resonance. Figures 1.7

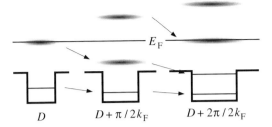

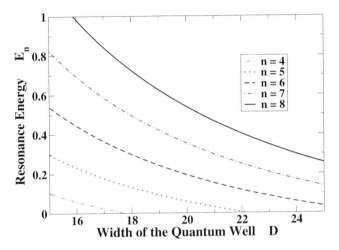

Figure 1.7. How resonances cross the Fermi level of a given multilayer system. Reproduced with kind permission from M.D. Stiles (Ref. [25]).

Figure 1.8. Resonances calculated for the free electron problem (well depth $= V = 0.25$ a.u.) discussed in text.

and 1.8 show these resonances and energies. In the latter figure, peak positions for several n values are shown as a function of the width (thickness) of the quantum well D, calculated using an incident plane wave representing a free electron.

In real magnetic multilayers, such peaks in the density of states can be observed using photoemission and inverse photoemission spectra; for related work see Refs. [36–39]. Photoemission, which is discussed in detail in a later chapter, is a technique in which photons of a particular energy, generally ultraviolet light or X-rays, are used to excite electrons in a given surface. The binding energy of the emitted electron in the solid state can be inferred from the photon energy and the kinetic energy emitted electron in the vacuum outside. Peaks are observed at energies corresponding to a large density of states in the material in the photoemission spectrum. Inverse photoemission can be described as the inverse of the above process, where electrons are forced to occupy the unoccupied states above the Fermi level; they emit photons to escape from such unstable states. The emitted photon energy carries information about the unoccupied level placement above the Fermi level. Photoemission probes the surface/subsurface region since the escape depth of the photoemitted electrons is of the order of a nanometer. In order to (experimentally) observe

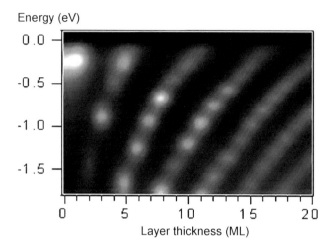

Figure 1.9.   A high resolution image of resonances observed in a photoemission experiment for Co/Cu. The bright spots identify integral numbers of monolayer thicknesses. Reproduced with kind permission from Y.Z. Wu [40].

the density of states peaks in the nonmagnetic spacer layer, the top magnetic layer needs to be stripped off. Hence, the quantum well states studied in photoemission may not be quite the same quantum well states that are present in a real magnetic multilayer. Nonetheless, there is a certain degree of correspondence between these states and the related states in magnetic multilayers.

Figure 1.8 is from a simple calculation, based on quantum wells and free electrons, which shows the resonances as a function of the width of the well. Figure 1.9 illustrates the photoemission intensity as a function of energy and spacer layer thickness. These figures show the fixed spacing between peaks as a function of thickness and variation of the peaks as a function of energy. There are some differences between what would be expected for a free electron model and what is observed in a real experiment. To gain more insight and see how the free electron model generalizes to real materials, it is instructive to rewrite the transmission probability in terms of the transmission probability for a step and the reflection amplitude at an isolated step $R_{step} = R = (\{k - q\}/\{k + q\})^2$ (from Equation (1.5)). This form (Equation (1.7)) clearly illustrates the contribution made by multiple reflection inside the quantum well. One round trip through the well has the amplitude $e^{i2qD}R$ from reflecting from each step and propagating in both directions through the well. The transmission probability is

$$T_{well} = \left| T_{step}e^{iqD}\frac{1}{1 - e^{i2qD}R} \right|^2 = \left| T_{step}e^{iqD}\sum_{n=0}^{\infty}(e^{i2qD}R)^n \right|^2. \tag{1.7}$$

The right hand side of Equation (1.7) (i.e., the sum of exponentials) shows explicitly the coherent multiple scattering in the well. The basic physics of quantum well states in real materials is captured by replacing the wave vector for propagating through the spacer layer, q, by the appropriate value from the real band structure and by replacing the reflection amplitude and transmission probability by the values calculated for a realistic

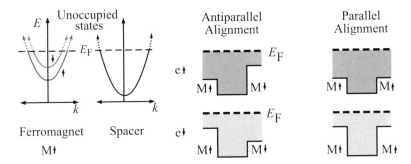

Figure 1.10. Spin polarized, free electron bands and quantum wells. Reproduced with kind permission from M.D. Stiles (Ref. [25]).

interface. When electrons encounter a ferromagnetic material, the potential step for the majority electrons will be different from that for the minority electrons. In a multilayer with two magnetic layers, there are four possible quantum wells formed depending on the relative alignment of the magnetizations (see Figure 1.10). The quantum well states for each of these are different because the potential steps, and hence reflection probability, are different for each quantum well. However, at a particular energy, such as the Fermi energy, the quantum well states in all of the wells have the same periodicity as a function of the thickness of the spacer layer, since the periodicity only depends on the wavelength of the electron in the spacer layer at that energy. These ideas are related to envelope function modulations of Bloch states found in the bulk. The so-called aliasing effect (discussed in Chapter 2), arising from the discrete Fourier sampling in the multilayer, are all closely tied to envelope modulations as well [37]. A more detailed discussion of the QW related experiments is given in Chapter 8.

1.4. Interlayer exchange coupling

The coupling between two magnetic layers depends on several factors, somewhat similar to the competition between various magnetic states. If a certain alignment is favored, the conclusion must be that it is energetically favorable compared to other possible alignments. The energy associated with these alignments, in real materials, are associated with the topology of the Fermi surfaces, i.e., as to how the states at various parts of the Fermi surface affect the above mentioned coupling. Sometimes, such details are not that straightforward to identify.

The interlayer exchange coupling may be written in terms of an energy that depends on the magnetization directions of the two layers, $\hat{\mathbf{m}}_i$, as

$$E = -JA\hat{\mathbf{m}}_1 \cdot \hat{\mathbf{m}}_2, \tag{1.8}$$

where A is the common area shared by the two layers. This coupling is also called bilinear coupling. Clearly, when $J < 0$, antiferromagnetic alignment of the two layers is favored. This exchange coupling constant can be evaluated using free electron (i.e., parabolic) bands for the magnetic as well as the spacer layers. Then the exchange coupling is simply the energy difference between parallel and antiparallel alignments shown in Figure 1.10.

This energy difference between parallel and anti-parallel alignments is approximately

$$J \approx \frac{\hbar v_F}{2\pi D} |R_\uparrow - R_\downarrow|^2 \cos(2k_F D + \phi), \tag{1.9}$$

for large spacer layer thicknesses, D. Here k_F, v_F are the Fermi wave vector and Fermi velocity in the spacer layer respectively while $R_\uparrow$ and $R_\downarrow$ are the reflection amplitudes for up and down spin reflection at the interface. The presence of the cosine term indicates that the exchange coupling can oscillate between positive and negative values, decays as D^{-1}, and the amplitude depends on the difference between up and down reflection coefficients. The latter clearly points to the fact that if there is no clear difference between spin polarization-driven effects, then there will not be any oscillations in the exchange coupling as discussed above!

The underlying physical reason for the oscillations is that the energies, at which the quantum well resonances occur, cross through the Fermi level as the thickness D of the spacer is varied (Figure 1.7). The resonances for each quantum well are different, but they possess the same period since these are tied to the Fermi level of the spacer, and not the magnetic properties of the other layers. The existence of a Fermi surface seems quite necessary for the observed oscillations in interlayer coupling and hence the GMR effect.

1.5.   3-dimensional model: Fermi surface nesting

The expression for interlayer coupling J, obtained previously, is for a simple one-dimensional QW. If the multilayer growth is coherent, then it is possible to obtain a straightforward expression for the 3-dimensional model. In this case, the QWs corresponding to different planar reciprocal lattice vectors $\mathbf{G}_\parallel$ contribute independently to the exchange coupling, since $G_\parallel$ is conserved at interface scattering events. Hence those contributions have to be integrated (or summed) over to obtain the total coupling. This is simply an integral of the one dimensional result (shown earlier for large D in Equation (1.9)) over the parallel reciprocal lattice vectors.

$$\frac{J}{A} \approx \frac{\hbar}{2\pi D} \int \frac{d^2 \mathbf{G}_\parallel}{(2\pi)^2} v_F(\mathbf{G}_\parallel)$$
$$\times \mathrm{Re}\big[\big(R_\uparrow(\mathbf{G}_\parallel) - R_\downarrow(\mathbf{G}_\parallel)\big)^2 \exp\big(i2k_z(\mathbf{G}_\parallel)D\big)\big] + \mathrm{O}\big(D^{-3}\big). \tag{1.10}$$

Now the contributions to this integral, from various parts of the Fermi surface, are likely to cancel due to the rapid oscillatory behavior of the exponential, unless its argument stays constant in some region of the relevant Fermi surface. Such behavior occurs at stationary points, called critical spanning vectors or nesting vectors. As the thickness D gets larger and larger, the Fermi surface region gets smaller and the oscillations die out. For free electrons, the critical spanning vector q_c is $2k_F$. In Figures 1.11 and 1.12 such critical spanning vectors are shown for free electrons as well as for several real materials. These critical spanning vectors can be tied to the observed long and short periods of oscillations in exchange coupling. However, to obtain a complete picture of such periods, the discrete multilayer structure has to be taken into account (i.e., a free electron type picture is insufficient – see the discussion of "aliasing" in Chapter 2).

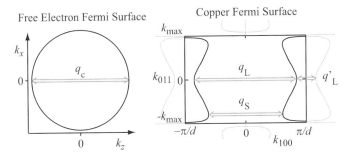

Figure 1.11. Critical spanning vectors for free electrons and those for Cu. In the latter case, note the long and short spanning vectors leading to long and short periods in interlayer exchange coupling, respectively. In order to understand this connection, one needs to invoke what is referred to as "aliasing" (see Chapter 2). Reproduced with kind permission from M.D. Stiles (Ref. [25]).

1.6. Strength of coupling: theory vs experiment

The coupling strength, J, shown in Equation (1.10), involves certain (spacer) Fermi surface related variables (for example, spanning vectors) in addition to interface related variables such as the reflection amplitudes. Now, experimentally, nesting vectors can be measured, but not the reflection amplitudes. In Figure 1.13, the comparison is not as perfect as one would like it to be. One reason being that the Fermi surfaces calculated using the best available theoretical techniques still are not accurate enough, due to insufficient treatment of many-electron effects which are responsible for magnetism among other things. Hence after a few oscillations, the calculated coupling is out of phase with the experimental coupling.

1.6.1. Growth related issues and measuring interlayer coupling

In this volume, pseudomorphic growth will be discussed in a separate chapter. When Fe is grown on fcc Cu(100) under appropriate conditions, we know that Fe takes the fcc structure of Cu, at least up to a few monolayers. This is an example of pseudomorphic growth. There are only a limited number of systems relevant to the GMR effect, such as Co/Cu or Fe/Cr, which exhibit pseudomorphic growth. Both these pairs have close lattice matches and hence can be grown on interfaces having different crystal orientations. This is, of course, a very attractive property for multilayer growth, although there is no guarantee of perfect interfaces etc. even with such matching, as discussed in a separate chapter. Apparently, there are no other known pairs of metals exhibiting the GMR effect and such similarities in lattice matching with various orientations. The other systems which come close to the above pair are Ag on Au or Fe, but only in the (001) orientation.

Although the physics responsible for the GMR effect is quite well understood, comparisons between theory and experiment have not be very successful, at least partly due to the above mentioned problems. Comparisons of experiments and theory are further complicated by the so-called "thickness fluctuations". The growth discussed above, even under almost ideal lattice matching, is never perfect. The substrate may not be perfectly flat and the growth may not occur layer-by-layer. However, if there are flat terraces large

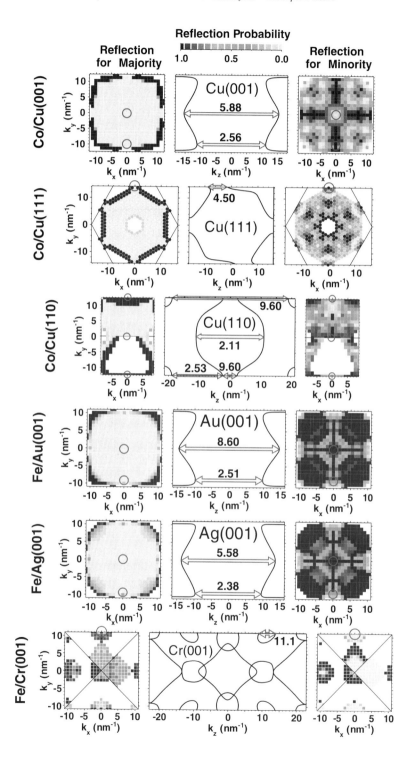

Figure 1.12. Spin-dependent, interface reflectivities and critical spanning vectors for several multilayer systems. The left and right panels show slices of the Fermi surfaces of the spacer material indicated in the middle panel based on the reflection probabilities from the magnetic layers for majority and minority spins. The shading scale is given at the top and the middle panel shows the critical spanning vectors, which are also indicated by circles in the left and right panels. Reproduced with kind permission from M.D. Stiles (Ref. [25]).

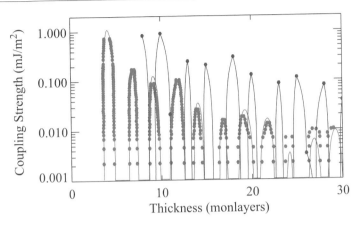

Figure 1.13. Calculated [41] and measured [42] coupling strengths for Fe/Au/Fe(100) multilayers. The thick (red) curve is the best fit to the experimental data (symbols). The thin (black) curve is a linear interpolation between the coupling strengths calculated (symbols) for complete layers. Reproduced with the permission from M.D. Stiles [25].

enough, the interlayer coupling can be averaged over an ideal thickness of n layers when comparisons are made, as in Figure 1.13. In addition, when the interlayer coupling varies rapidly with the layer thickness, it is very likely that the measured coupling is affected by the thickness fluctuations. If such fluctuations reduce the measured interlayer coupling significantly, it will be quite difficult to compare theoretical and experimental couplings. Some of these problems can be solved by maintaining a layer-by-layer growth with a narrow growth front. However, energetics and temperature (thermodynamics) will also play key roles here, in determining whether there is island formation. Higher temperatures will reduce island formation; unfortunately, higher temperatures can also give rise to undesirable effects, such as interdiffusion.

We will be discussing interdiffusion at a metallic interface in a separate chapter. In the Fe/Cr system, interdiffusion at the interface appears to be responsible for the reversal of sign in the GMR effect. Certain impurities in the multilayer system NiCr/Cu/Co/Cu (see below) have also been detected to reverse the sign in the GMR effect. One significant advance in the measuring of oscillation periods is the use of wedge-samples (discussed in a separate chapter). For such samples, variations in thickness can be monitored under the same growth conditions as well as the same substrate.

1.6.2. GMR in magnetic sensors

One other important application of GMR is in sensors that can detect magnetic fields. When such a device is placed in a magnetic field, the electrical resistance of that device

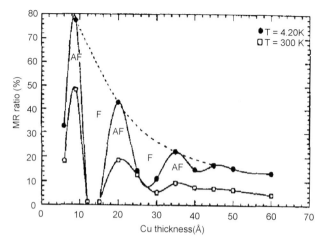

Figure 1.14.   Observed GMR ratios of Co(1.5 nm)/Cu multilayers as a function of the thickness of Cu layers at $T = 4.2$ K (closed circles) and $T = 300$ K (open squares), from Ref. [30]; solid lines are guides to the eye. Reproduced with permission from Elsevier.

should be sensitive to the external magnetic field. In typical magnetic multilayers with a nonmagnetic spacer layer, the fields required to saturate the change in resistance are relatively high. In order to use these in (GMR) sensors or other practical devices, it is essential for them to be sensitive to reasonably small changes in the field strength.

As discussed earlier, the magnetic coupling between layers occurs through the nonmagnetic spacer layer and this coupling strength J oscillates between ferromagnetic (F) and antiferromagnetic (AF) coupling with increasing (nonmagnetic) spacer layer thickness as seen in Figure 1.14. It was realized that using this oscillatory behavior, these systems can be "spin engineered" so that with an external applied field, spin arrangement could be switched from F to AF, thereby altering the electrical resistance (see Figure 1.14). This measurement can be tied to the coupling strength, which was also found to have certain systematic variations with spacer layers formed from $3d$, $4d$ and $5d$ transition metals. As for technological applications, Ru was found to have several desirable properties. Films as thin as 2 to 3 Å were found to produce AF coupling in addition to their thermal stability and growth properties (Refs. [18,26]).

In order to detect small magnetic fields in magnetic read heads, the GMR spin-valve has turned out be quite useful. The simplest form of this device is a sandwich structure having two ferromagnetic layers separated by a thin Cu layer. The magnetics of AMR as well as GMR sensors is often tied to a phenomenon called "exchange biasing", which will be discussed in Chapter 4 of this book.

1.6.3.   Biquadratic coupling

Since the exchange coupling J varies from F to AF (from positive to negative) with the thickness of the spacer layer, one can choose a thickness which makes the exchange coupling quite weak. At such thicknesses (called a node in exchange coupling), another form of coupling, termed "biquadratic coupling", takes over. This coupling, first observed in

Fe/Cr/Fe [27], forces the magnetic moments in adjacent layers to lie perpendicular to one another. To lowest order, the intralayer exchange coupling helps to keep the magnetization uniform in each layer. However, to next order, there can be fluctuations in magnetization in a given layer, around its average value. The contribution to energy at this level is quadratic in both $\hat{\mathbf{m}}_1$ and $\hat{\mathbf{m}}_2$, resulting in the so-called biquadratic coupling energy (in contrast to bilinear coupling discussed earlier)

$$E = -J_2(\hat{\mathbf{m}}_1 \cdot \hat{\mathbf{m}}_2)^2. \tag{1.11}$$

Measured values of J_2 are always negative indicating that such magnetizations prefer to be perpendicular to one another. As Slonczewski (Ref. [28]) has shown, biquadratic coupling has an extrinsic origin due to disorder, unlike (intrinsic) bilinear coupling. A review of biquadratic interlayer coupling can be found in Ref. [29].

1.7.   Selected multilayer systems

Among the large numbers of multilayer systems studied, two GMR systems stand out with regard to their significance. Those are; Co/Cu which is regarded as the prototypical system and Fe/Cr where the GMR effect can be very large at low temperature. Table 1.1 shows some of the experimental GMR ratios for these systems and a metal alloy system.

1.7.1.   Co/Cu(001)

The two metals Co and Cu are well lattice-matched and Co/Cu(001) turns out to be the most extensively studied multilayer system, both theoretically and experimentally (Ref. [1]). The exchange coupling involves two periodicities. One period is long and can be linked to the belly of the free electron like Fermi surface at the zone center $\bar{\Gamma}$. The other period is short and is associated with the neck of the Fermi surface (see Figure 1.11). Calculations indicate that for the long period, the reflection amplitudes for both spins are rather small, for thick Co layers. However, as the Co layer thickness is decreased, the reflection amplitude for the minority spins increase rapidly at **k**-points close to the zone center. The reflection amplitudes appear to be quite sensitive to the Co layer thickness.

For the short period, there is a gap in the Co minority states with the same symmetry as the Fermi surface electrons of Cu around critical **k**-point. This gap is quite narrow in

Table 1.1   Comparison of GMR ratios in selected multilayer systems

System	Ratio max. $\Delta\rho/\rho\%$	Temperature (K)	Reference
Co/Cu(001)	50	300	[30]
	80	4.2	[30]
Fe/Cr(001)	150	4.2	[31]
	30	~300	[31]
FeNi(12 nm)/Cu(4 nm)	68	77	[34]
	78	4.2	[34]

energy and hence the reflectivity is strongly energy dependent. This energy dependence gives rise to a large reduction of the coupling strengths in thin Co/Cu(001) when compared with infinitely thick Co layers. The experimentally measured short period, 2.6 ML, is in good agreement with theoretical predictions of the critical spanning vector while the long period measured from several experiments ranges from 5.9 to 8.0 ML (see Refs. [17,32,33]).

1.7.2.   Multilayers grown on Fe whiskers

Multilayers grown on Fe whisker substrates can be controlled to a high atomic scale precision, removing atomic scale disorder. Experimental results from such multilayer systems can be meaningfully compared with theoretical results where ideal atomic structures are almost always assumed and employed. The Fe whiskers are among the most perfect metallic crystals and near perfect (100) surfaces can be obtained by in situ ion sputtering and thermal annealing [44]. Wedge shaped interlayers (see Chapter 8), which make it feasible to study varying interlayer thicknesses, were grown on the Fe whisker substrate, for several interlayer elements such as Cr, Ag, Au, Mn, V, Cu and Al. In addition, on top of the interlayer, a thin Fe layer was deposited. RHEED and SEMPA (see Figure 1.15) scans were used to monitor the oscillations in exchange coupling. There appears to be remarkable agreement between theory and experiment for oscillation periods of the metals listed above, illustrated in Table 1.2, except for Vanadium (not shown in the table).

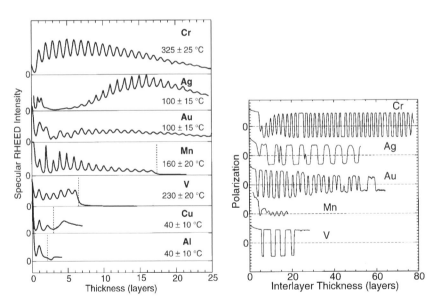

Figure 1.15. (Left) Reflection High Energy Electron Diffraction (RHEED) intensity oscillations as a function of thickness for deposition of various metals on to the Fe substrate at the indicated temperature; (Right) Polarization line scans using Scanning Electron Microscopy with Polarization Analysis (SEMPA) showing oscillations in the top Fe layer magnetization as a function of the thickness of the spacer. Used with the permission of Elsevier from Ref. [44].

Table 1.2   Comparison of measured and calculated periods for coupling along the (001) direction for various interlayers sandwiched between a thin Fe layer and a Fe whisker substrate

Interlayer	Measured periods (ML) (Unguris *et al.* [44])	Theory (ML) (Stiles [43])	Theory (ML) (Bruno *et al.* [17])
Cr (0.144 nm/layer)	2.105 ± 0.005	2.10	
	12.1 ± 1	11.1	
Ag (0.204 nm/layer)	2.37 ± 0.07	2.45	2.38
	5.73 ± 0.05	6.08	5.58
Au (0.204 nm/layer)	2.48 ± 0.05	2.50	2.51
	8.6 ± 0.3	9.36	8.60

Reproduced from Ref. [44] with the permission of Elsevier.

1.7.3.   Fe/Cr system

Fe/Cr system, which exhibits a rich, spin density wave (SDW) character, has received significant attention since 1986 when a trilayer Fe/Cr/Fe system showed antiferromagnetically coupled Fe layers on either side of a thin Cr film [9]. In the following years, there were numerous further studies of this system, such as the ones in Refs. [13] and [31]. However, there have been various controversies surrounding this system, some attributable to poor sample quality, in the early days of GMR work. These early experiments revealed only a single (long) period of oscillation (see Ref. [45] for a detailed review). After a short period was found, there was the important question as to whether the SDWs were necessary for the short period oscillations seen in (interlayer) exchange coupling. However, when this coupling was found to exist above the bulk Néel temperature of Cr and other multilayers were also found with short period oscillations, the interest in the SDWs seems to have tapered off.

The role of SDWs in the exchange coupling of layers has remained illusive and uncertain, while collinear and well as non-collinear SDWs have been observed in this system from Neutron Scattering experiments. In bulk Cr, 3 types of SDWs, labeled C, I, H, have been found. Interlayer exchange coupling mediated by a Cr spacer layer has some unique features since Cr can exist in various magnetic states. This coupling depends on whether Cr exists in one of its SDW forms or in a nonmagnetic (paramagnetic) form. The roughness distribution also affects the magnetic coupling. The trilayers Fe/Cr/Fe(001) grown on Fe whiskers is the best understood case. The experimental results are consistent with a CSDW up to a certain Cr thickness Figure 1.15. For more details on this particular system, see Refs. [44,45] and references therein. In addition, a brief theoretical discussion of the SDWs will be given in Chapter 2.

1.7.4.   Multilayered alloys and nanowires

Slater–Pauling curve [48] (Figure 1.16) shows how the saturation magnetization of various alloys of magnetic elements varies along the 3*d* row. The right half of the curve, which mainly consists of Ni-based fcc alloys, forms a well defined straight line with a

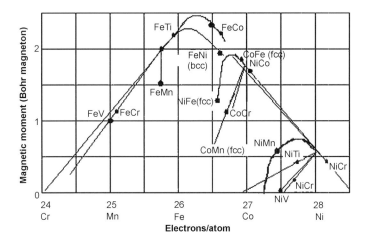

Figure 1.16.   A schematic of the so-called Slater–Pauling curve (from Ref. [48] with permission from Elsevier). This plot represents the experimental values of the saturation magnetization of Fe-, Ni-, and Co-based alloys vs average number of electrons.

slope of −1. The left half consists mostly of Fe-based bcc alloys. The maximum moment occurs for $Fe_{0.7}Co_{0.3}$. It is energetically favorable to keep the majority d band full irrespective of the alloy composition along the main branches of this curve. A brief look at the Slater–Pauling curve is sufficient to convince us that various alloys of $3d$ magnetic elements are also likely to be magnetic, Permalloy, which has a 80:20 Ni:Fe ratio with random (or almost random) site occupations, being one such example. This is a ferromagnetic alloy of Fe and Ni with a Curie temperature of about 560 °C. Here also we can expect strong spin dependent scattering as discussed above.

Alloying is one way to test the dependence of the GMR oscillation periods on the spacer material. Alloy concentration in the spacer layers can change the band structure, Fermi surface, and the critical spanning vectors. When the Cu spacer is alloyed with small amounts of Ni, it remains nonmagnetic but the Fermi surface contracts. For (111) and (110) oriented multilayers, Ni doping gives rise to an increase in the oscillation period of the long period while for (001) multilayers, it yields a decrease. Some of the above mentioned changes have been observed to be consistent with the Fermi surface properties [49,51]. For example, in $Co/Cu_{1-x}Ni_x(110)$ system, the increase in the long-period of oscillation with increasing Ni concentration was directly tied to the neck orbit of the Cu-Ni (spacer) Fermi surface [51].

Magnetic nanowires can be fabricated by various methods, such as using electrodeposition into templates (see Figure 1.17). Various groups have demonstrated that multilayer wires can be made out of Co/Cu, NiFe/Cu, CoNi/Cu, Ni/Cu and Fe/Cu. Some of these systems, such as Co/Cu or permalloy/Cu, are known to exhibit single-crystal structures. Multilayered nanowires have been useful in testing some of the CPP (current perpendicular to plane)-GMR predictions, determining spin diffusion lengths as well as studying temperature dependence of the CPP-GMR effect [34]. In addition, in current induced switching (or reversal) of magnetizations (see Chapter 7), such wires can be used. Nanowires are ideal for studying high injection density currents in order to probe the changes in spin

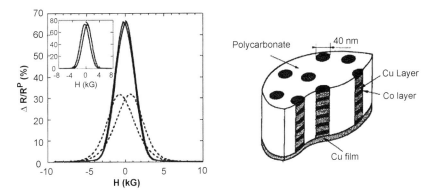

Figure 1.17. CPP Magnetoresistance vs applied field parallel to the layers at 77 K for permalloy(12 nm)/ Cu(4 nm) (solid line) and Co(10 nm)/Cu(5 nm) (dashed line) multilayered nanowires. The inset shows the same for permalloy/Cu sample at 4.2 K (from Refs. [34,35]). Also shown is an array of multilayered nanowires in a nanoporus, track-etched polymer membrane. Reproduced with permission from Elsevier.

configurations of multilayers. For perpendicular recording media and tunneling magnetoresistance devices, these wires can be quite useful.

1.8. Scattering of electrons: a simple picture

GMR effect is directly tied to spin dependent scattering of electrons, which can depend on a number of factors such as the density of states of a transition metal (Figure 1.18). The fact that spin dependent scattering is important was recognized early on by Mott [7]. His straightforward but insightful description is based on a two-current model for the two spin directions, i.e., the spin transport was perceived as being due to two almost independent spin channels. In a paramagnetic metal such as Pd, these different spins should experience an identical resistance that is quite different from the resistance experienced by different spins in a ferromagnetic metal. The following is a brief description of Mott's early work: Consider a ferromagnetic d band metal having a magnetization $M(T)$ at temperature T. Let z be the fractional magnetization with respect to its saturation value, i.e., $z = M(T)/M_0$ where M_0 is the saturated magnetization at zero temperature. Then at temperature T, a fraction, $(1 - z)/2$ of the unoccupied d states will have their spins parallel to M and a fraction $(1 + z)/2$ will have their spins antiparallel. Now writing $N_1(E)$ as the density of states of spins parallel to M and $N_2(E)$ as the density of states antiparallel to M, using a parabolic form for the density of states, we have

$$N_1(E) = C\sqrt{(E_1 - E)} \tag{1.12}$$

and

$$N_2(E) = C\sqrt{(E_2 - E)}, \tag{1.13}$$

with E_1 and E_2 being the highest energies that the two spins can have. Clearly, E_1 and E_2 are different when there is a nonzero spin splitting, but will be identical $(= E_0$ say) above the Curie temperature. The corresponding energy dependent relaxation times, τ_1

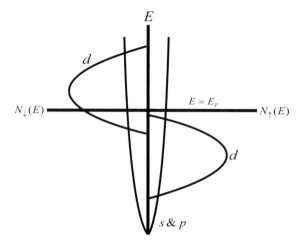

Figure 1.18.   Schematic density of states representing s, p and d states of a transition metal.

and τ_2, can be related to these densities of states as,

$$1/\tau_1 \propto N_1(E) + \gamma \tag{1.14}$$

and

$$1/\tau_2 \propto N_2(E) + \gamma, \tag{1.15}$$

where γ is a contribution from spin independent scattering.

$$1/\tau_1 = \text{const}\left(T/M_I\theta^2\right)\left\{\frac{\sqrt{(E_1 - E)}}{\sqrt{(E_0 - \zeta_0)}} + \alpha\right\} \tag{1.16}$$

with $\alpha \simeq 1/4$ and θ being the Debye temperature. A similar formula can be obtained for τ_2.

If ζ_0' represents the energy of the highest occupied state (Fermi energy) at $T = 0$ when the states are split and ζ_0 when $z = 0$, then

$$\sqrt{\frac{E_1 - \zeta_0'}{E_0 - \zeta_0}} = (1 - z)^{1/3}, \qquad \sqrt{\frac{E_2 - \zeta_0'}{E_0 - \zeta_0}} = (1 + z)^{1/3} \tag{1.17}$$

since $N_d(E)$ is proportional to $n_0^{1/3}$.

Now the conductivity σ may be written as

$$\sigma = -Ne^2/m \int \frac{(\tau_1 + \tau_2)}{2} \frac{\partial f}{\partial E} \, dE, \tag{1.18}$$

using a Drude type formula, but with energy dependence of the relaxation time included. If the partial derivative in the integrand is taken to be none zero only at $E = \zeta_0'$, then the resistivity $\rho = 1/\sigma$ is given by,

$$\rho(z, T) = \text{const}\left(T/m\theta^2\right)\left\{\frac{1}{(1 - z)^{1/3} + \alpha} + \frac{1}{(1 + z)^{1/3} + \alpha}\right\}^{-1}. \tag{1.19}$$

This expression is a function of two variables, T and z, and when the resistances between the saturated case ($z = 1$) and the paramagnetic case are compared we obtain,

$$\frac{\rho(z = 1, T)}{\rho(z = 0, T)} \simeq 0.34, \tag{1.20}$$

when $\alpha \simeq 0.25$.

From this rather rudimentary description, it is very easy to see that the magnetization has a strong effect on the resistance and that when a given spin is parallel to a given magnetization, it experiences a lower resistance compared to what an antiparallel spin would experience.

The magnetoresistance ratio (MRR), is defined as

$$\frac{R_{AP} - R_P}{R_P} \tag{1.21}$$

where R_{AP} and R_P refer to the resistances with antiparallel and parallel alignments of spins relative to a given magnetic layer (respectively). If we denote the resistances for up and down channels as $R^{\uparrow}$ and $R^{\downarrow}$, then

$$R_P = \left(1/R^{\uparrow} + 1/R^{\downarrow}\right)^{-1} \tag{1.22}$$

and

$$R_{AP} = \left(R^{\uparrow} + R^{\downarrow}\right)/4 \tag{1.23}$$

when the mean free path of the electrons is much higher than the repeat length of the multilayers (Ref. [46]). Using the asymmetry parameter in GMR, $\alpha = R^{\downarrow}/R^{\uparrow}$ or $\beta = (-R^{\downarrow} + R^{\uparrow})/(R^{\downarrow} + R^{\uparrow})$ the MRR, defined above, can be written as,

$$\frac{(1 - \alpha)^2}{4\alpha} \tag{1.24}$$

from which it follows that $R_{AP} > R_P$, giving a highly simplified explanation of the higher resistance encountered in the antiparallel case.

This mechanism is illustrated in the following figure (Figure 1.19) for $\alpha > 1$, i.e., when the resistivity is smaller for the majority spins. In the parallel (P) configuration, spin up

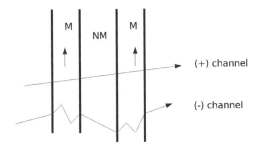

Figure 1.19.   Schematic diagram showing GMR scattering (following Ref. [50], used with permission from Elsevier). In the parallel (P) configuration shown, majority electrons in the (+) channel experience little or no resistance and hence a short circuit effect occurs. In the antiparallel (AP) configuration (not shown), electrons in the (+) and (−) channels will experience a significant resistance when going through the slab with opposite magnetization, with no short circuit effect.

electrons are the majority in both magnetic slabs and are weakly scattered giving rise to a resistance r, smaller than the resistance R encountered in the spin down channel. The short-circuit effect caused by the low resistance (fast electrons) makes the resistivity low in the P configuration (see Ref. [47] and Figure 1.19).

We can try to understand the above results in terms of the density of states (DOS) of transition metals, Ni, Co and Fe, provided we ignore all other scattering mechanisms except for the electron–electron scattering. Although at various interfaces the features of DOS at the Fermi level are likely to change, we first focus on bulk $3d$ metals listed above. It is also important to note that in a solid, there is $s - p - d$ hybridization and that there are no pure d states, but mostly d-like, p-like and s-like states. However, in Mott's simple picture, the $3d$ states are assumed to be responsible for spin dependent scattering. In bcc Fe, the DOS of majority and minority states at the Fermi level are not very different, while in Co and Ni, the majority spin DOS is not that high compared to the minority spin DOS. This implies that incoming minority spins are likely get scattered more, although the d electrons do not participate directly in the transport process, but act as scattering centers.

Impurity Effect or the extrinsic potential: One of the most dramatic examples due to impurities on the GMR effect has been seen in NiCr/Cu/Co/Cu multilayers where impurities are added in the NiCr layer to reverse its spin asymmetry. The asymmetry parameter, β defined earlier, is negative for large thicknesses of NiCr and has a negative effect on the GMR effect of Cu/Co/Cu. An extensive study (Ref. [52]) with various impurities has revealed a close connection between the asymmetry parameter β and the Slater–Pauling curve shown in Figure 1.16. All the negative β elements and alloys lie on the positive slope side of the Slater–Pauling curve while the positive β materials correspond to the negative slope region.

Diffusive scattering: If the interfaces are rough (i.e., interface region does not consist of a sharp boundary separating element A from element B), one might expect spin independent scattering. This is most likely due to averaging of spin up and down potentials in a region of space, where there is no clear and well defined spin sensitivity. Again in the early days of the GMR discoveries, such effects were not that well recognized.

1.9.   Magnetic tunnel junctions

By 2006, GMR-based read heads began to be replaced by read heads based on another megnetoresistance phenomenon associated with tunnel junctions. For completeness, we introduce magnetic tunnel junctions (MTJs) here. In Chapter 7, a more detailed description of these tunnel junctions can be found. A MTJ is similar to the spin valves discussed earlier with one important difference. This difference has to do with replacing the non-magnetic and metallic spacer layer with a thin insulating tunneling barrier. Tunneling magnetoresistance (TMR) values much higher than 100% have been observed at room temperature in devices that use certain insulating materials. These are labeled CPP (or current perpendicular to plane) devices. The perpendicular current flow through a tunnel junction device is ideal for ultra high density magnetic recording since the device can be directly attached to magnetic shields which can then be used as contacts. In contrast, in many GMR devices where the current flow is parallel to the layers (CIP – current in plane), the sensor has to be electrically isolated from the conducting magnetic shields.

As discussed in this volume (see Chapter 7), MRAM devices consist of magnetic storage cells or bits, where each bit is a thin film multilayered structure that has two magnetic states. These two states are designed to have different magnetoresistances. For example, a small magnetic field will switch the its resistance from state "0" to state "1". By making use of the GMR effect, it is possible to get large signals in response to small applied fields. Moreover, the signal can be read nondestructively, i.e., without changing its magnetic state. Such a (non-volatile) device has low power consumption in addition its speed advantage, since a bit does not have to be written every time it is read. However, one problem with GMR devices is their conductive nature, which makes it necessary to wire them in series and hence the actual signal detectable gets scaled by N, the number of GMR cells. A key advantage in a MTJ over a CIP GMR device is its CPP nature. The electrical contacts thus occupy the same space as the MTJ device, making the cell quite small. CIP and CPP geometries have different scaling lengths associated with spin dependent scattering (mean free path vs spin diffusion length). (See Ref. [18] and references therein for more details.)

1.10. Half-metallic systems

Up to now, we have been considering magnetic elements which are metallic, with spin up and spin down electrons contributing to the Fermi level density of states. In the search for highly spin polarized systems, one could envision having a situation where one spin channel shows metallicity while the other spins are completely insulating. Such materials, where tunneling currents show 100% spin polarization when used in tunneling magnetoresistance devices, have been found and are called half-metals (for example, see Ref. [53]). In Chapter 9 of this volume, half-metallic systems are discussed in some detail with emphasis on specific materials.

1.11. Summary

In this chapter, a straightforward introduction to the GMR effect was provided. Some applications, such as magnetic recording, sensors, and multilayered nanowires were discussed briefly. A simple picture of how the magnetic exchange interaction gets transmitted through the spacer layer was provided using quantum well ideas from elementary quantum mechanics. In addition, various related experimental systems were discussed and compared to theoretical calculations whenever possible. There is ample evidence that the periods associated with oscillatory exchange-coupling are closely tied to the thickness of the spacer layers in ideal (or near perfect) multilayer systems. However, such periods of oscillations appear to be relatively insensitive to disorder; the same does not hold for the strength of coupling. Also, note that with increasing disorder, the oscillations in exchange coupling begin to fade away.

More details on the topics and techniques discussed in this chapter can be found in the following chapters as well as in numerous review articles and books written since the discovery of the GMR effect. Chapter 8 contains more information on the quantum well structures. Volume 200 of the *Journal of Magnetism and Magnetic Materials*, the series *Ultrathin Magnetic Structures I–IV* (see Ref. [54] for Vol. II), and Ref. [55] constitute a small sample of such references. In this chapter, our focus has been centered on transition metal multilayers; rare-earth systems are discussed in Refs. [56,57].

Acknowledgements

The author is indebted to Dr. Mark D. Stiles for providing numerous relevant material for this chapter. He is also grateful to Drs. S.S.P. Parkin, Z.Q. Qiu and Y.Z. Wu for useful discussions and granting permission to use relevant material.

References

[1] S.S.P. Parkin, X. Jian, C. Kaiser, A. Panchula, K. Roche, M. Samant, Proceedings of the IEEE **91**, 661 (2003).

[2] S.S.P. Parkin, in: B.W. Wessels (Ed.), in: Annual Review of Materials Science, vol. 25, Annual Reviews Inc., Palo Alto, CA, p. 357 (1995).

[3] M.H. Kryder, R.W. Gustafson, J. Magn. Magn. Mat. **287**, 449 (2005).

[4] H.J. Richter, et al., IEEE Trans. Magn. **42** (10), 2255 (2006).

[5] K.G. Ashar, Magnetic Disk Drive Technology: Heads, Media, Channel, Interfaces and Integration (IEEE Press, New York, 1997).

[6] T.R. McGuire, R.I. Potter, IEEE Trans. Magn. **Mag-11**, 1018 (1975).

[7] N. Mott, Proc. Roy. Soc. **156**, 368 (1936).

[8] A. Fert, I.A. Campbell, Phys. Rev. Lett. **21**, 1190 (1968).

[9] P. Grünberg, R. Schreiber, Y. Pang, M. Brodsky, H. Sowers, Phys. Rev. Lett. **57**, 2442 (1986).

[10] M.B. Salamon, S. Sinha, J.J. Rhyne, J.E. Cunningham, R.W. Erwin, J. Borchers, C.P. Flynn, Phys. Rev. Lett. **56**, 259 (1986).

[11] C.F. Maykrzak, J.W. Cable, J. Kwo, M. Hong, D.B. McWhan, Y. Yafet, J.V. Waszcak, C. Vettier, Phys. Rev. Lett. **56**, 2700 (1986).

[12] C. Carbone, S.F. Alvarado, Phys. Rev. B **36**, 2433 (1987).

[13] M.N. Baibich, J.M. Broto, A. Fert, F. Nguyen Van Dau, F. Petroff, P. Etienne, G. Creuzet, A. Friederich, J. Chazelas, Phys. Rev. Lett. **61**, 2472 (1988).

[14] G. Binasch, P. Grunberg, F. Saurenbach, W. Zinn, Phys. Rev. B **39**, 4828 (1989).

[15] S.S.P. Parkin, N. More, K.P. Roche, Phys. Rev. Lett. **64**, 2304 (1990).

[16] M.A. Ruderman, C. Kittel, Phys. Rev. **96**, 99 (1954);
T. Kasuya, Prog. Theor. Phys. **16**, 45 (1956), 58;
K. Yosida, Phys. Rev. **106**, 893 (1957).

[17] P. Bruno, C. Chappert, Phys. Rev. Lett. **67**, 1602 (1991);
P. Bruno, C. Chappert, Phys. Rev. B **46**, 261 (1992);
P. Bruno, C. Chappert, in: R.F.C. Farrow, et al. (Eds.), Magnetism and Structure in Systems of Reduced Dimension, Plenum Press, New York (1993).

[18] S.S.P. Parkin, Phys. Rev. Lett. **67**, 3598 (1991).

[19] F. Petroff, A. Barthélémy, D.H. Mosca, D.K. Lottis, A. Fert, P.A. Schroeder, W.P. Pratt, R. Loloee, S. Lequien, Phys. Rev. B **44**, 5355 (1991).

[20] J. Fassbender, F. Norteman, R.L. Stamps, R.E. Camley, B. Hillebrands, G. Guntherodt, Phys. Rev. B **46**, 5810 (1992).

[21] K. Ounadjela, D. Miller, A. Dinia, A. Arbaoui, P. Panissod, G. Suran, Phys. Rev. B **45**, 7768 (1992).

[22] Y. Huai, R.W. Cochrane, J. Appl. Phys. **72**, 2523 (1992).

[23] J. Zhang, P.M. Levy, A. Fert, Phys. Rev. B **45**, 8689 (1992).

[24] P.J.H. Bloeman, W.J.M. de Jonge, R. Coehoorn, J. Magn. Magn. Mater. **121**, 306 (1993).

[25] M.D. Stiles, private communication;
M.D. Stiles, J. Appl. Phys. **79**, 5805 (1996);
M.D. Stiles, J. Magn. Magn. Mater. **200**, 322 (1999).

[26] S.S.P. Parkin, D. Mauri, Phys. Rev. B **44**, 7131 (1991).

[27] M. Ruührig, et al., Phys. Stat. Solidi A, Applied Research **125**, 3172 (1991).

[28] J.C. Slonczewski, Phys. Rev. Lett. **67**, 3172 (1991);
J.C. Slonczewski, J. Magn. Magn. Mater. **150**, 13 (1995).

[29] S.O. Demokritov, J. Phys. D **31**, 925 (1998).

[30] D.H. Mosca, F. Petroff, A. Fert, P.A. Schroeder, W.P. Pratt Jr., R. Laloee, J. Magn. Magn. Mater. **94**, L1 (1991).

[31] E.E. Fullerton, M.J. Conover, J.E. Mattson, C.H. Sowers, S.D. Bader, Phys. Rev. B **48**, 15755 (1993).

[32] M.T. Johnson, S.T. Purcell, N.W.E. McGee, R. Coehoorn, J. aan de Stegge, W. Hoving, Phys. Rev. Lett. **68**, 2688 (1992).

[33] W. Weber, R. Allenpach, A. Bischof, Europhys. Lett. **31**, 491 (1995).

[34] A. Fert, L. Piraux, J. Magn. Magn. Mater. **200**, 338 (1999).

[35] S. Dubois, et al., J. Magn. Magn. Mater. **165**, 30 (1997).

[36] Z.Q. Qiu, N.V. Smith, J. Phys. Condens. Matter **14**, R169 (2002).

[37] F.J. Himpsel, Phys. Rev. B **44**, 5966 (1991);
J.E. Ortega, F.J. Himpsel, Phys. Rev. Lett. **69**, 844 (1992);
J.E. Ortega, F.J. Himpsel, G.J. Mankey, R.F. Willis, Phys. Rev. B **47**, 1540 (1993).

[38] N.B. Brookes, Y. Chang, P.D. Johnson, Phys. Rev. Lett. **67**, 354 (1991).

[39] R.K. Kawakami, E. Rotenberg, E.J. Escorcia-Aparicio, H.J. Choi, J.H. Wolfe, N.V. Smith, Z.Q. Qiu, Phys. Rev. Lett. **82**, 4098 (1999).

[40] Y. Wu, private communication.

[41] J. Opitz, P. Zahn, J. Binder, I. Mertig, Phys. Rev. B **6309**, 4418 (2001).

[42] J. Unguris, R.J. Celotta, D.T. Pierce, Phys. Rev. Lett. **79**, 2734 (1997).

[43] M.D. Stiles, Phys. Rev. B **48**, 7238 (1993);
M.D. Stiles, Phys. Rev. B **54**, 14679 (1996).

[44] J. Unguris, R.J. Cellota, D.A. Tulchinsky, D.T. Price, J. Magn. Magn. Mater. **198–199**, 396 (1999).

[45] R.S. Fishman, Spin-density waves in Fe/Cr trilayers and multilayers, J. Phys.: Condens. Matter **13**, R235 (2001).

[46] A. Fert, Mat. Sc. Forum **59–60**, 439–480 (1990).

[47] A. Barthélémy, et al., J. Magn. Magn. Mater. **242–245**, 68–76 (2002).

[48] P.H. Dederichs, R. Zeller, H. Akai, H. Ebert, J. Magn. Magn. Mater. **100**, 241 (1991).

[49] S.S.P. Parkin, C. Chappert, F. Herman, Europhys. Lett. **24**, 71 (1993).

[50] A. Fert, P. Grünberg, A. Barthélémey, F. Petroff, W. Zinn, J. Magn. Magn. Mater. **140–144**, 1 (1995).

[51] S.N. Okuno, K. Inomata, Phys. Rev. Lett. **70**, 1711 (1993).

[52] C. Vouille, A. Barthélémy, F. Elokan Mpondo, A. Fert, Phys. Rev. B **60**, 6710 (1999).

[53] J.M.D. Coey, M. Venkatesan, J. Appl. Phys. **91**, 8345 (2002).

[54] B. Heinrich, J.A.C. Bland (Eds.), Ultrathin Magnetic Structures, vol. II, Springer, Berlin (1994).

[55] L.H. Bennet, R.E. Watson (Eds.), Magnetic Multilayers, World Scientific, Singapore (1994).

[56] C.F. Majkrzack, J. Kwo, M. Hong, Y. Yafet, D. Gibbs, C.L. Chien, J. Bohr, Adv. Phys. **40**, 99 (1991).

[57] J.J. Rhyne, R.W. Erwin, in: K.H.J. Buschow (Ed.), Magnetic Materials, vol. 8, North-Holland, Elsevier, Amsterdam (1995).

Chapter 2

Overview of First Principles Theory: Metallic Films

Simple metallicity is commonly understood as the ease with which a given material is able to conduct electricity or heat. Hence this very common understanding of a simple metal is directly tied to its transport properties. However, transport in general is a nontrivial phenomenon that can depend on the type of material under observation. For example, electronic transport may depend on many factors such as the electronic structure, lattice vibrations (phonons), defect structure and related scattering mechanisms. In this volume, special attention will be paid to electronic transport which depends on magnetic interactions. Drude is credited with providing a simple (stochastic) model of metallic transport, based on kinetic theory, around the beginning of the 20th century (see, for example, Ashcroft and Mermin [1]). This was merely a few years after the electron was discovered by J.J. Thomson. Using assumptions similar to those that were used in the kinetic theory of gases, Drude managed to obtain a simple DC conductivity formula, which reads,

$$\sigma = \frac{ne^2\tau}{m}. \qquad (2.1)$$

Here n is the number density of electrons, m, e are the charge and mass of an electron respectively while τ is a relaxation time. A basic assumption in Drude's theory is that a solid is a collection of ions and electrons (where the ions are stationary while the electrons are free to move around). The collisions in the Drude model were assumed to be due to electrons scattering off ions. A relaxation time τ (i.e., time till next collision) played an important role in this stochastic model. Drude theory was used to estimate other transport coefficients such as thermal conductivity and thermopower, but clearly had difficulty in several situations.

Going beyond the Drude theory required the introduction of quantum mechanics and the Pauli principle to describe electrons in metals. Quantum mechanical effects, such as exchange (spin–spin) interactions, form an essential part of our understanding of metals today. To a large extent, normal metals can be understood in terms of an effective one particle theory. Electrons occupy one-particle (or quasi one-particle) levels according to the Pauli principle when in their ground state. The Fermi surface defines a constant energy surface in **k**-space that corresponds to occupied states with the highest energy (Fermi energy, ϵ_F). In metals at low temperature, there is a sharp cut-off in occupancy at the Fermi surface. This sharp cut-off is almost always responsible for various oscillatory features seen in metals, as discussed later. Such Fermi surface effects, responsible for several phenomena discussed in this volume, can be quite accurately calculated and understood nowadays using first principles band structure. Hence, a somewhat detailed discussion of band calculations, especially pertaining to thin films, is carried out in the following sections. Even in the presence of magnetic interactions, which are truly many-body effects, band calculations have been successful when dealing with the so-called itinerant magnets. This is basically due to the successes of the mean-field approaches, such as the density functional theory (DFT) based approximations. When highly local-ized magnetic interactions are present, one-electron theory based methods may not work. An often quoted example is when an odd number of (unit cell) electrons have to occupy a given number of (spin-degenerate) bands, there will be a half-filled band which implies metallicity. Although, by doubling the unit cell (due to antiferromagnetic or some other symmetry breaking mechanism) one can overcome the above, there are many examples of such materials (such as NiO) that are non-metallic and show non-band-like behavior. In such cases, model Hamiltonians such as the Spin Hamiltonian, Heisenberg, Anderson or Hubbard have been useful. However, for understanding the GMR effect and many other phenomena tied to metallicity, band theory has played a sufficient and crucial role and for that reason, we review some of the modern band structure methods.

2.1.   First principles band structure

Theoretical methods that are used to calculate (energy) band structures have come a long way since the original formulation of quantum mechanics. From those early attempts to calculate band structures by Slater and Krutter, Slater, Bardeen, Wigner and Seitz to some of the modern band structure methods such as the Full Potential Linear Augmented Plane Wave (FPLAPW) method and others, a great deal of scientific knowledge has been gathered. There is a vast literature on such attempts that we cannot try to cover here; instead, the salient features of first principles methods that have been most successful in describing (itinerant) magnetic behavior in solids (including multilayers) are discussed briefly. There are two important aspects that have to be addressed with regard to first principles band structure and these are,

- reduction of the many-electron problem and,
- solving the effective one-electron problem.

Once the original many-electron problem is reduced to an effective one-electron prob-lem (to be discussed later), the band structures can be calculated from first principles (i.e., without effectively using experimental input) to almost machine precision using numer-ical techniques, employing high quality basis sets. The basis set is used to expand the eigenstates of the Hamiltonian. The expansion coefficients, when found self-consistently determine the eigenstate while the eigenvalues are referred to as the band structure. Plane waves are probably the most analytically convenient basis set to use. However, they do not mimic the behavior of the atomic states in the core regions of the atom; hence, an exceedingly large number is required in order to expand such states. There are modern pseudopotential methods, that have been developed over the past two decades, that ad-dress some of the above issues. However, we will not be discussing such methods in this brief review.

Before the linear methods (discussed below) came into play, there were essentially two types of approaches that were utilized. One of them had to do with a fixed basis set (such

as LCAO-linear combinations of atomic orbitals or plane waves) while the other type re-
lied on matching partial waves (KKR or APW). A brief discussion of their strengths and
weaknesses will be given below. The KKR-Green Function method [2] is one of the old-
est and elegant methods for studying electronic structure of localized defects embedded
in an ideal, infinite host crystal. Starting from the Green function of the ideal crystal, the
multiple scattering of the electrons at the defect potentials is evaluated exactly. For metals, the KKR-Green function method can be considered optimal. The first and foremost
advantage of this method is the small, but optimal basis set which is both energy and site
dependent.

In the LCAO (or tight-binding) method, the wave function $\psi_{\mathbf{k}}(\mathbf{r})$ can be expanded as
follows:

$$\psi_{\mathbf{k}}(\mathbf{r}) = \sum_N A_N^{\mathbf{k}} \sum_{\mathbf{R}} \Xi_N(\mathbf{r} - \mathbf{R}) \exp(i\mathbf{k} \cdot \mathbf{R}). \tag{2.2}$$

Here $\mathbf{R}$ refers to lattice vectors and the sum over those assures that the wave function
contains the effects due to the translational symmetry of the lattice, i.e., it satisfies the
Bloch's theorem. The lattice sum (over $\mathbf{R}$), in principle, involves all the direct lattice
vectors. The basis functions are denoted by Ξ_N; if these are fixed in the sphere regions
centered around nuclei and atomic-like, it may lead to errors due to the incompleteness
of the basis set. A better basis set would correspond to obtaining such functions from the
self-consistent radial potential in the solid, as is the case with augmented methods. Over
the years, these ideas have led to substantial improvements in efficiency and quality of the
band structure calculations.

2.1.1.   *Linearization in energy*

Linear methods came of age during the 1970s and O.K. Andersen is credited with the
development of their essential aspects [3]. The theory underlying linearization has to do
with using a radial wave function (ϕ) and its energy derivative ($\dot{\phi}$) to represent the radial
dependence of the basis function inside the sphere regions. The introduction of ϕ and $\dot{\phi}$
naturally ensured the orthogonality to core states while accomplishing the linearization
in energy, which resulted in a vast improvement in efficiency without loss of significant
accuracy. All the eigenvalues at a given $\mathbf{k}$-point could now be obtained in one diagonalization, instead of the more time consuming search for individual eigenvalues in the
nonlinear problem, as in the APW method [13].

Accurate, linearized band structure calculations began to be reported during the mid
to late 1970s [4–7] and were applied to study thin metallic films during the same period.
Since the linearization proved to be quite efficient as well as accurate, it has been adopted
by the modern band calculation methods such as LAPW [7], LMTO [8,9] or LASTO [10],
to name some. The LMTO method has its roots in the original KKR technique, while
LASTO method uses Slater orbitals to define the interstitial basis functions.

Although there were some efforts to treat semi-infinite solids, the most successful surface calculations were carried out with slab geometries (see Figure 2.1). Typically, a finite
number of layers (preferably more than five) is used to model a slab assuming that it is
thick enough to reproduce bulk (electronic structure) behavior in the central (inner) layers.
The slab geometry assumes 2-dimensional periodicity parallel to the surface, and treats

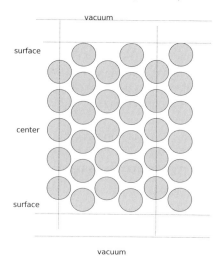

Figure 2.1.   Slab geometry.

the vacuum region quite accurately. The physics behind such calculations is that in transition metals, the screening length of the electrons is comparable to the interlayer spacing. Other important features of these thin slab approximations are discussed in Appelbaum and Hamann [11].

2.1.2.   Basis sets for thin metallic films

The band structure problem has to do with solving a Schrödinger equation self-consistently. To do this in practice, we need to have (1) an accurate one particle potential that represents the physical system (for example, a thin film) and (2) the ability to calculate the wave functions. As discussed earlier, the wave function is determined through an expansion in terms of basis functions. In most popular band structure methods, the basis set plays a crucial role. Construction of these basis sets, for example, for thin films, has to be carried out mimicking the behavior of the crystal potential in various regions of space (see Figure 2.2). Here, space is divided into sphere (around nuclei), interstitial and vacuum regions. The lowering of symmetry due to the presence of a surface (or an interface) has to be carefully dealt with. Hence, unlike in some bulk band structure applications, the so-called muffin-tin approximation, where the potential is taken to have a rather simple behavior (such as a constant in the interstitial), is likely to breakdown.

For example, one of the most popular methods is the Linearized Augmented Plane Wave (LAPW) method. It combines the advantages of energy independent, muffin-tin Hamiltonian methods [5,12] with the matrix element determination of the original APW method [13]. This also facilitates the inclusion of the "full potential" where all the non-spherical pieces of the one particle potential are taken into account without any shape approximations [14]. In the LAPW method, the plane wave refers to the basis set in the interstitial region while it is augmented into the spheres using radial solutions of the semi-relativistic Schrödinger equation. So, inside the sphere, the basis set is generated self-consistently in response to the (actual) radial potential. Such a basis function at a **k**-point

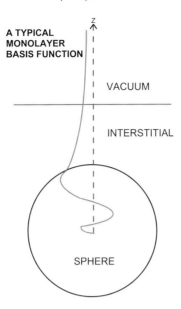

A TYPICAL
MONOLAYER
BASIS FUNCTION

VACUUM

INTERSTITIAL

SPHERE

Figure 2.2.   A part of a typical basis function for a monolayer film, showing its behavior in the three regions
(sphere, interstitial and vacuum) of space, along the normal direction to the slab.

$\mathbf{k}_{\parallel}$ of the 2-dimensional Brillouin zone may be written as,

$$\Xi_{\mathbf{G}_{\parallel},G_{\perp}}^{\mathbf{k}_{\parallel}}(\mathbf{r}) = C \exp\big(i(\mathbf{k}_{\parallel} + \mathbf{G}_{\parallel}) \cdot \mathbf{r}_{\parallel}\big) \exp(i G_{\perp} r_{\perp}). \tag{2.3}$$

Note that the vector $\mathbf{G} = (\mathbf{G}_{\parallel}, G_{\perp})$, where $\mathbf{G}_{\parallel}$ is a reciprocal lattice vector corresponding to the 2-dimensional, periodic lattice. For a film geometry, the translation symmetry of the two-dimensional crystal can be exploited using (planar) reciprocal lattice vectors in order to Fourier expand the charge density and the potential. The perpendicular component $G_{\perp}$ is defined in terms of an artificial repeat length along the z direction (i.e., there is an implicit supercell here). The basis function inside the ith sphere takes the form

$$\Xi_{\mathbf{G}_{\parallel},G_{\perp}}^{\mathbf{k}_{\parallel}}(\mathbf{r}) = \sum_{L} \big[\alpha_{L}(\mathbf{G}_{\parallel}, G_{\perp})\phi_{l}(r : E_{l}) + \beta_{L}(\mathbf{G}_{\parallel}, G_{\perp})\dot{\phi}_{l}(r : E_{l})\big] Y_{L}(\hat{\mathbf{r}}), \tag{2.4}$$

where ϕ_l and $\dot{\phi}_l$ (energy derivative) are solutions arising from the radial, scalar relativistic Schrödinger equation corresponding the actual radial potential, at a given energy E_l. The linearization attempts to approximate the behavior of the energy dependence of the radial function in an adequate region of energy for a given l value (for example, for $l = 2$, it would attempt to cover the d band width). The structure constants α_L and β_L are needed for the proper handling of the lattice sum in Equation (2.2).

However, the basis has to be defined in the vacuum regions as well. This is accomplished again using the actual, planar averaged potential in the vacuum region in a Schrödinger equation. The resulting solutions constitute the z dependent part of the basis function,

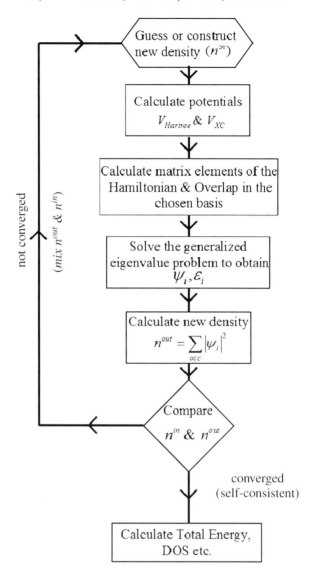

Figure 2.3.   A flowchart showing some of the essential steps of a modern self-consistent band structure calculation.

$$\Xi^{\mathbf{k}_\parallel}_{\mathbf{G}_\parallel, G_\perp}(\mathbf{r}) = \exp\left(i(\mathbf{G}_\parallel + \mathbf{k}_\parallel) \cdot \mathbf{r}_\parallel\right)\left[A(\mathbf{G}_\parallel, G_\perp)u(z : E_\nu)\right.$$
$$\left. + B(\mathbf{G}_\parallel, G_\perp)\dot{u}(z : E_\nu)\right], \qquad (2.5)$$

so that the function is connected smoothly to the interstitial basis function using the coefficients A and B. Here the z dependent functions u and $\dot{u}$ are obtained by solving a one-dimensional Schrödinger-like equation for the planar averaged potential in the vacuum region at a given energy parameter E_ν.

Once the basis set is constructed, matrix elements

$$H_{ij} = \langle \Xi_i | H | \Xi_j \rangle$$

and

$$O_{ij} = \langle \Xi_i | \Xi_j \rangle$$

(with i or j denoting the basis indices, instead of $\mathbf{G}_\|, G_\perp$) have to be evaluated at the appropriate $\mathbf{k}$-point. Then a generalized eigenvalue problem

$$\sum_j \hat{H}_{ij} A_j = \lambda_i \sum_j \hat{O}_{ij} A_j,$$

has to be solved in order to obtain the eigenstates $\{A_j\}_{j=1,n}$ and eigenvalue λ_i for the ith state at the given $\mathbf{k}$-point in the Brillouin zone. These eigenvalues, obtained self-consistently as illustrated in Figure 2.3 as a function of the $\mathbf{k}_\|$ vector, constitute the band structure of the thin film.

2.1.3. *Full potential*

Accurate Density Functional Theory based all electron methods, where the full charge density of a given system is utilized to calculate potentials, depend on accurate and practical ways to evaluate the direct Coulomb potential. An elegant treatment, discussed by Weinert [14] and Hamann [15], paved the way to solve the Poisson equation for a given charge density without any shape approximations (such as in the muffin-tin approximation) to the potentials. This also avoids the use of Ewald-like methods. The basic idea is that the potential outside a closed region of charge density depends on that charge density only through the multipole moments of the charge distribution in the closed regions. Hence, to evaluate the potential in the interstitial region, one needs to know the charge density in the interstitial and the multipole moments of the charge distribution in different sphere regions. However, the multipole moments do not uniquely determine the charge distribution. This allows one to replace the rapidly varying charge density inside the spheres by a smooth (pseudo) charge distribution having identical multipole moments. This latter distribution can be constructed so that it has a rapidly converging Fourier transformation. The following radial function, as suggested in Ref. [14], can be used to construct a pseudodensity satisfying the necessary requirements;

$$n_{lm}^{pseudo}(r) = a_{lm}(r/R_0)^l \left[1 - (r/R_0)^2\right]^p, \quad r \leqslant R_0, \tag{2.6}$$

where R_0 is a sphere radius and l, m are angular momentum labels. Here a_{lm} should be chosen to have correct multipole moments and p is arbitrary.

2.1.4. *Some aspects of group theory*

Physics is a study of symmetries and the theory of groups, developed in abstract mathematics, has been quite useful in the studies of symmetry, to say the least. For example, the Bloch's theorem, which was essential to reduce the infinite lattice problem, exploits the translational symmetry of a Bravais lattice. Operations corresponding to translations by lattice vectors, form an Abelian (commutative) group. The symmetry group of a crystal consists of all the operations that leave the crystal invariant. These operations are not simply translations, but could be a combination of translations and rotations in general. The set of all such operations forms a group.

The definition of an abstract group consists of three conditions. A non-empty set G together with a binary operation $\odot$ (an operation that acts between two elements of G resulting in an element in G) is called a group $\{G, \odot\}$ if and only if,

1. $(a \odot b) \odot c = a \odot (b \odot c)$;
2. There exists an identity element e, such that $a \odot e = e \odot a = a$ for every $a \in G$;
3. Given $a \in G$, there exists an element $b \in G$ such that $a \odot b = b \odot a = e$.

Note that the operation $\odot$ need not be commutative and in condensed matter physics, we encounter many such non-Abelian groups, such as some of the 32 point groups which describe local symmetries compatible with 3-dimensional Bravais lattices. Matrix representations of abstract groups are quite important and useful in describing various symmetries. The set of all $n \times n$ unitary matrices with their determinants equal to $+1$ forms a group labeled $SU(n)$. If the entries are real (i.e., matrices orthogonal) then they constitute a subgroup of $SU(n)$ of proper rotations, labeled $SO(n)$. A subgroup is a group in its own right under the same binary operation, but is a subset of the original group. All the proper (determinant $= 1$) and improper (determinant $= -1$) rotations in n dimensions constitute the group $O(n)$; $SO(n)$, for example, is a subgroup of $O(n)$. However, symmetry operations of an infinite crystal (either 2 or 3 dimensional) consist of rotations as well as translations and are usually expressed as $\{R|\mathbf{t}\}$ where $R \in O(3)$ or $O(2)$ with $\mathbf{t}$ representing a translation. When the symmetry operations of the crystal cannot be separated into rotations and translations (i.e., there are symmetry elements $\{R|\mathbf{t}\}$ where $\mathbf{t}$ is not a Bravais lattice vector) the symmetry group is called non-symmorpic; otherwise the group is symmorphic.

The symmetries of the Hamiltonian lead to degenerate energy eigenstates. These symmetries could be exploited to construct and reduce the Brillouin zone, into an irreducibe one (the size of the Brillouin zone would be reduced by a factor corresponding to the number of symmetry operations in the "factor group" G/T where T denotes the subgroup of lattice translations of the full symmetry group G). When dealing with symmetries of eigenstates, it is not necessary to know all the symmetry operations of the crystal; i.e., there is a further simplification, namely these symmetry groups can be mapped on to matrix representations. The theory of such matrix representations provides us a way to label eigenstates (and energy bands). A matrix representation is a homomorphic mapping of a group into a set of matrices (i.e., it is not necessarily a one to one mapping). Symmetry labels such as Γ, Δ that are used in band theory refer to matrix representations that are irreducible (usually referred to as *irreps*). For two-dimensional band structures, it is customory to use labels such as $\bar{\Gamma}$ or $\bar{\Delta}$.

As an example, let us consider the point group C_{4v} which can be used to describe the 4-fold, square symmetry about a lattice site with some additional mirror planes. An atom on a (100) surface of a bcc or fcc slab (Figure 2.1) has this local symmetry; in addition, C_{4v} is the point group symmetry of the eigenstates at $\bar{\Gamma}$ (see Figure 2.4) in the Brillouin zone for the band structure of such a slab. This point group has 8 elements which

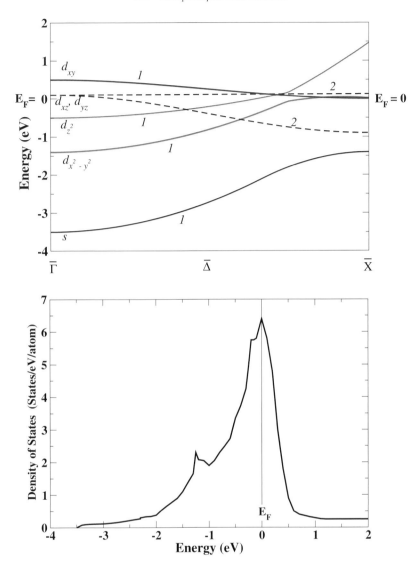

Figure 2.4.   Unpolarized Fe monolayer bands along $\bar{\Delta}$ and the corresponding density of states. Note how the $x - y - z$ (bulk) degeneracy at the zone center is broken since the normal (z) direction samples a different potential compared with the planar ($x - y$) directions. The labels 1 and 2 in the top figure refer to z reflection symmetry.

can be divided into 5 different (mutually exclusive) conjugate classes; the elements in a conjugate class are related to one another by a unitary transformation. It can be shown that the number of conjugate classes is exactly equal to the number of *irreps*. Table 2.1 illustrates what is called a character table where the character of an element A in an *irrep* is defined to be Trace(A). All the necessary symmetries that distinguish eigenstates at $\bar{\Gamma}$, as mentioned above, are now listed in Table 2.1 in a compact form. In passing, we note

Table 2.1   Character table of C_{4v}

C_{4v}	I	C_2	$2C_4$	$2\sigma_v$	$2\sigma_d$	orbitals
A_1	1	1	1	1	1	s, d_{z^2}
A_2	1	1	1	-1	-1	
B_1	1	1	-1	1	-1	$d_{x^2-y^2}$
B_2	1	1	-1	-1	1	d_{xy}
E	2	-2	0	0	0	d_{xz}, d_{yz}

The first row specifies the symmetry operations I (identity), 2-fold (C_2) and 4-fold (C_4) rotations as well as the mirror plane operations σ_v and σ_d. The corresponding d orbitals and their symmetries are also shown.

that in order to deal with magnetic band structures, the so-called double groups have to be utilized. For a comprehensive review of the group theory concepts discussed here, the reader is referred to Ref. [16].

One more useful application of group theory is in symmetrizing observables such as the charge density or one particle potential. For example, when the charge density is calculated from an irreducible (wedge of the) Brillouin zone (IBZ), it usually does not reflect the lattice compatible space group symmetry. In order to carry out symmetrization, the appropriate space group operators, $S = \{R|\mathbf{t}\}$ where R is a rotation and $\mathbf{t}$ is a translation, have to be used. Due to crystal symmetry, one can restrict the (wavefunction product, $\psi_{\mathbf{k}}\psi_{\mathbf{k}}^*$, based) summations to only the IBZ, leading to

$$\sum_{\mathbf{k}\in BZ} \psi_{\mathbf{k}}(\mathbf{r})^* \psi_{\mathbf{k}}(\mathbf{r}) = \sum_{\substack{\mathbf{k}\in IBZ \\ S}} \left[S\psi_{\mathbf{k}}(\mathbf{r}) \right]^* S\psi_{\mathbf{k}}(\mathbf{r}), \tag{2.7}$$

and a significant saving of computational time. The sum over S in the above equation is a projection operator that projects out the part of $\psi_{\mathbf{k}}\psi_{\mathbf{k}}^*$ that transforms according to the fully symmetric representation of the space group.

A similar symmetrization can be used to generate the so-called "lattice harmonics". These are linear combinations of spherical harmonics that are compatible with the local point group symmetry $\mathcal{G}$. Starting with a given spherical harmonic $Y_{l_v m_v}$ and summing over point group operations $R \in \mathcal{G}$ yields the (normalized) lattice harmonic K_{l_v}, i.e.,

$$K_{l_v}(\hat{\mathbf{r}}) = \frac{1}{C} \sum_{R\in\mathcal{G}} R Y_{l_v m_v}(\hat{\mathbf{r}}) \tag{2.8}$$

with C being a normalization constant. When rotating functions as in Equation (2.7) or Equation (2.8), one can use the definition $S\psi(\mathbf{r}) = \psi(S^{-1}\mathbf{r})$.

2.1.5.   Simple examples

We will illustrate several pedagogical points with a Fe monolayer calculation employing a LAPW basis set. The energy bands along one high symmetry direction of the 2-dimensional Brillouin Zone are shown in Figure 2.4. Note how the symmetry breaking due to the vacuum (z points to the vacuum, while $x - y$ axes lie in the plane of the

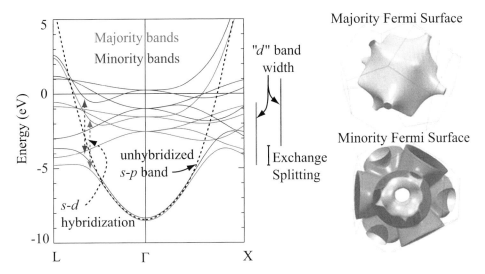

Figure 2.5. Band structure of spin polarized cobalt. Note the exchange-split d bands and unsplit $s - p$ bands. Reproduced with permission from M.D. Stiles (private communication).

monolayer) partially lifts the degeneracies of the t_{2g} $3d$ states d_{xy} and d_{xz}, d_{yz} at $\bar{\Gamma}$, the center of the zone. Also note how the so-called e_g states ($d_{x^2-y^2}$ and d_{z^2}) are split at $\bar{\Gamma}$, due to different symmetries along the z direction (perpendicular to the film) compared to x and y directions. In general, at an interface, symmetry broken states are to be expected. A related (schematic) density of states (shown in Figure 2.4) illustrates the band narrowing due to having just a single monolayer of Fe. Fermi surfaces of thin slabs will undoubtedly show finite size effects and these will affect transport as well as other Fermi surface driven phenomena. The Fe-monolayer bands in Figure 2.4 were calculated without spin polarization and hence does not show any spin splitting of the five d bands.

In Figure 2.5, spin polarized bands for bulk Co are shown. Here the exchange split d bands are clearly seen. Note how the $s - p$ bands, which do not contribute significantly to magnetism, do not show any spin splitting. The exchange splitting of the d bands is smaller than the d band width but still is sufficiently large to produce interesting magnetic phenomena, as well as different Fermi surfaces (Figure 2.5).

2.1.6. Spectral representation

The density of states (DOS) is a very useful way to get some handle of the 'spaghetti' that is usually called the 'band structure'. However, one must remember that the DOS is an integrated form of the band structure and hence, invariably some information will be lost due to integration. However, there is more transparency; i.e., one can look at site-projected or l-projected DOS in order to understand what changes occur where etc.

One particle Green function $G(\mathbf{r}, \mathbf{r}'; E)$ is usually expressed in terms of the eigenfunctions ψ_ν as,

$$G(\mathbf{r}, \mathbf{r}'; E) = \sum_\nu \frac{\psi_\nu(\mathbf{r})\psi_\nu^*(\mathbf{r}')}{E + i\epsilon - E_\nu} \quad \epsilon \longrightarrow 0+. \tag{2.9}$$

Here $E_\nu s$ are the eigenvalues representing the band structure and the integral,

$$n(E) = -\frac{1}{\pi} \int_V d\mathbf{r} \operatorname{Im} G(\mathbf{r}, \mathbf{r}; E) \tag{2.10}$$

yields the local DOS (see Figure 2.5). Although there are other ways to express the DOS, this form appears to be more transparent. If the wavefunction ψ_ν is written as a site projected or a l projected sum, identifying those particular projections and carrying out the above integral, restricted to such terms, will yield a site projected or a l projected DOS.

2.2. Density functional theory: reduction of the many-electron problem

An appropriate basis set provides a basic platform to get some of the essential elements of the problem of multilayering under control. However, the reduction of the many-electron problem is also necessary in order to compute the band structure. One of the popular (and computationally feasible) techniques, namely the density functional theory based the local density approximation, will be discussed briefly in this section. In Chapter 6, a more rigorous discussion of the theory will be given. Here, one can question the use of the term "first principles" used to describe such calculations as being disingenuous to some extent. However, currently, the above approach is regarded as by far the best possible way to approximately describe the time independent electronic structure problem using minimal experimental input.

Density functional theory (DFT) has been the basis of a large number of calculations of electronic structure over the last few decades. Although such ideas based on the density did exist early on, such as in the Thomas–Fermi model, modern density functional theory is attributed to the work of Hohenberg and Kohn, who proved two important theorems [17]. First, properties of the ground state of an inhomogeneous, interacting electron gas are universal functionals of the electron density, $n(\mathbf{r})$. Second, the ground state energy functional, $E\{n\}$, is stationary with respect to variations in the electron density and attains its minimum value $E\{n_0\}$ at the true ground state density, n_0: i.e.

$$E\{\psi\} = \langle \psi | H | \psi \rangle \geqslant E\{\psi_0\} = E\{n_0\}. \tag{2.11}$$

Later Kohn and Sham, made these ideas practical by connecting the interacting system with a noninteracting fictitious system [18]. Since there are plenty of elegantly written review articles (for example, Ref. [20]) on DFT, I shall not go in to a detailed discussion here. However, a brief description is in order due its ubiquitous nature where electronic structure calculations are concerned. The idea behind DFT is to come up with some external potential, when combined with the (known) classical Coulomb (Hartree) interactions, that will yield the correct ground state density for an interacting electron gas. If this is possible, then it is easy to come up with a "noninteracting electron" (in fact, a dressed electron) problem which will lead to an effective single particle problem. The above mentioned external potential is universally known as the "exchange–correlation potential".

2.3. Itinerant magnetism

In real materials, electrons interact with one another as well as the ions and possibly other external sources. This many ($N \approx 10^{23}$) electron problem is clearly quite nontrivial although the N-electron wavefunction contains all the relevant information for ground state properties. Hence, various models and approximations are employed to carry out practical calculations. One of the most important and practical approximations is the so-called effective one electron approximation, which leads to the standard band theory. This is obviously a mean-field approach and in metallic and other condensed matter systems, a large number of equilibrium properties can be reliably calculated in this way.

A reasonably simple model for magnetic coupling in transition metals can be built from an effective one electron approximation to the electronic structure, although the electronic structure of transition metal ferromagnets is quite complicated. Ferromagnetism in transition metals is driven by atomic-like exchange and correlation effects in the partially filled d-electron shells. The atomic-like effects suggest a localized description of this part of the electronic structure. However, the d orbitals are strongly hybridized with other d orbitals in neighboring atoms and also with the $s - p$ orbitals. The strong hybridization suggests an itinerant description of the electronic structure. Circumstances under which these methods are applicable and related aspects of the physics constitute an ongoing area of research, and the resulting models are not simple [19]. Simplifying the models requires approximations that favor one aspect of the physics over the other. Here we discuss models favoring the itinerant aspects because the interlayer exchange coupling depends strongly on the properties of the electrons at the Fermi surface and a realistic description of the Fermi surface requires treating the itinerant nature of the d electrons.

The local-density approximation (LDA) [18] and its spin-polarized version [20] (LSDA) accurately describe the itinerant aspects of the electronic structure while treating the atomic-like exchange and correlation effects in a mean-field theory. That is, all of the complicated electron–electron interactions are lumped into a local potential that depends on the local density [21]. This approximation was derived for computing the ground state properties of materials. For transition metal ferromagnets, it works quite accurately for properties such as the cohesive energy, equilibrium volumes, and the magnetic moment [24]. In Chapter 6, a formal discussion of its applicability is given, in terms of a rigorous, effective action approach.

Itinerant magnetism is understood in terms of band theory. The ferromagnetic $3d$ metals, such as Fe, Co and Ni, are considered to be good examples of itinerant magnets. Here band splitting near the Fermi level, into majority and minority spin states (see, for example, Figure 2.5), is responsible for the stabilization of the ferromagnetic state. Stoner criterion, which is discussed below, is usually satisfied when the above situation occurs. Another feature unique to an itinerant picture is the presence of interstitial regions with negative spin density in such materials. The spin density calculated in Ref. [22] for a ferromagnetic bcc Fe slab (Figure 2.6) shows negative and positive spin density regions (as experimentally verified), which supports the itinerant nature of magnetism in bcc Fe. Fermi surface instabilities are also responsible for other forms of magnetism (such as antiferromagnetism) within the band picture.

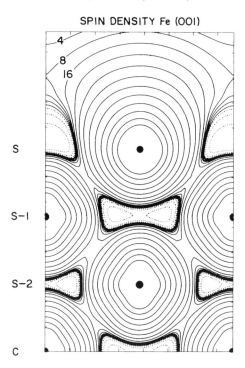

SPIN DENSITY Fe (OOl)

Figure 2.6.   Spin density of a 7 layer bcc Fe(001) slab from Ref. [22], reproduced with the permission from the American Physical Society. Note the positive and negative spin density regions resulting from an itinerant approach to magnetism.

2.4.   Localized vs. itinerant magnetic moments

The opposite extreme of the band model is based on interactions among localized magnetic moments. This form is usually described by phenomenological models, such as the Heisenberg model. In this model, two spins, s_i and s_j, interact with each other with an interaction energy $J_{ij}s_i \cdot s_j$ where J_{ij} is the exchange constant. Another phenomenological model, namely the Hubbard model in one dimension, predicts ferromagnetism when the on-site Coulomb interaction is sufficiently large with one hole-off half filling (Nagaoka theorem); There are no known exact solutions to the Hubbard Hamiltonian in higher dimensions, except for small clusters (see Chapter 10).

Both models (itinerant and localized) are quantum mechanical in origin; the exchange interaction between spins being primarily responsible for the magnetic order. Although one can try to relate the itinerant and localized models of magnetism, there is strictly no direct, one to one mapping between the two. For example, an itinerant band model would predict bcc Fe (ground state) moments to be about 2.12 μ_B per Fe atom with negative (minority) spin densities in the interstitial (between atom) regions [22]. A localized spin Hamiltonian will clearly not yield such predictions. We will later discuss these issues in some detail.

An approximate and simple way to describe the exchange (and related) effects, following Ref. [23], is as follows. Pauli principle prevents the occupation of the same orbital by two spins that are parallel. Hence the effective (Coulomb) interaction between parallel spins is different from that of antiparallel spins. In rigorous treatments of this effect, it is customary to define an exchange–correlation hole, around a given spin, where similar spins are absent. Depending on the strength of this effective Coulomb interaction, it may be (energetically) favorable to have parallel alignment of spins in some spatial region.

The direct Coulomb (i.e., Hartree) and exchange interactions can be written in a mean-field approximation as,

$$H_I = I \sum_{i\sigma} \left\{ \langle c_{j\bar{\sigma}}^+ c_{j\bar{\sigma}} \rangle c_{i\sigma}^+ c_{i\sigma} - \langle c_{j\sigma}^+ c_{j\bar{\sigma}} \rangle c_{i\bar{\sigma}}^+ c_{i\sigma} \right\} \qquad (2.12)$$

with $\bar{\sigma}$ denoting a spin opposite to σ. The creation operator, $c_{i\sigma}^+$, can be thought of as creating a Wannier state [1] at site i with spin σ. Also, in a periodic lattice, there is a well known result connecting the Wannier state with a sum of Bloch states (with Bloch vectors **k**) as,

$$c_{i\sigma} = \frac{1}{\sqrt{N}} \sum_{\mathbf{k}} a_{\mathbf{k}\sigma} e^{-i\mathbf{k}\cdot\mathbf{R}_i}, \qquad (2.13)$$

where $\mathbf{R}_i$ is a Bravais lattice vector.

The interaction term H_I can now be written in terms of the Bloch states which will bring out the explicit, long range magnetic interactions that can exist in a periodic lattice, such as ferromagnetic as well as antiferromagnetic. The (simplest) antiferromagnetism that is implicit here, results from a change in symmetry (doubling the unit cell) and could be tied to a $\mathbf{Q} = \mathbf{G}/2$, **G** being a reciprocal lattice vector with **Q** describing the variation of spin order. The above arguments point to expectation values such as,

$$n_\sigma^{\mathbf{Q}} = \sum_{\mathbf{k}} \langle a_{\mathbf{k}+\mathbf{Q}\sigma}^+ a_{\mathbf{k}\sigma} \rangle \qquad (2.14)$$

which support simple antiferromagnetic interactions tied to a critical spanning vector **Q**, which is also referred to as Fermi surface nesting.

2.5. Stoner criteria

To discuss the Stoner criteria for ferromagnetism briefly, one begins with a nonmagnetic density of states. If this density of states is high enough (to be defined) at the Fermi level, then it will be energetically favorable to occupy (empty) spin-up states above the Fermi level, at the expense of (occupied) spin down states. A one particle Hamiltonian can be solved easily to yield

$$E_{k\sigma} = \epsilon_k + I n_{\bar{\sigma}} \qquad (2.15)$$

where ϵ_k represents the nonmagnetic band states while $E_{k\sigma}$ identifies the spin-split bands. Note that $\bar{\sigma}$ denotes a spin that is opposite to a given spin σ and I, the so-called Stoner parameter. The above equation reduces to the following self-consistency condition involving a band splitting, Δ, given by

$$\Delta = E_{k,\uparrow} - E_{k,\downarrow} = I\mu(\Delta) \tag{2.16}$$

where $\mu(\Delta)$ is the induced local moment due to band splitting. If this condition holds self-consistently, then the system favors ferromagnetism.

Although the simple Stoner picture is based on a rigid band shift, as illustrated above, it can be easily incorporated into a self-consistent band calculation, with a Stoner parameter I.

Ferromagnets can either be saturated (maximum possible magnetization achieved) or unsaturated. There are two central aspects of the interactions that have to be clarified in order to understand saturated ferromagnets. First, there has to be ferromagnetic coupling among sites in a given region and second, at a given site, the intrasite interactions should favor saturation. Both these effects should coexist in order to have a strong ferromagnet. Several many-particle effects may be responsible for the (possible) appearance of strong ferromagnetism in hard magnets based on the $3d$ metals, the first being a local Hund's-rule type correlation effect providing the maximum local magnetization due to exchange splitting in the $3d$ shells. The second is an enhanced polarization effect similar to Stoner mechanism of saturated ferromagnetism in itinerant systems. This is often thought of as arising from local d vs. non-d exchange contributions, due to spin polarizations induced by the magnetic d electrons.

The exchange interactions determine the Curie temperature of a ferromagnetic metal; this is the temperature below which ferromagnetism develops and is of the order of a tenth of an electron volt (or less) for most ferromagnetic metals. However, this does not provide all the necessary information about magnetic interactions. A more realistic description has to include other (smaller) magnetic energies (such as magnetic anisotropies) in order to determine the alignment of spins with respect to the lattice. The associated energies are several orders of magnitude smaller than the exchange energies; such energies, though not capable of producing direct magnetic order, are relevant when the direction of polarization is important (see Chapter 4), as in various magnetic read/write devices.

2.6.   RKKY theory and interlayer coupling

RKKY (Ruderman–Kittel–Kasuya–Yoshida) Theory is based on the free electron gas and the exchange coupling arising from it. The indirect magnetic interaction between two spins mediated by free electrons is the RKKY interaction [25]. In the free electron model, the exchange coupling between two spins, separated by $\mathbf{R}$, is given by [26],

$$J(\mathbf{R}) \approx (\text{constant}) F(2k_F R) \tag{2.17}$$

where

$$F(x) = \left(x\cos(x) - \sin(x)\right)/x^4. \tag{2.18}$$

The constant in Equation (2.17) depends on the individual spins and the Fermi wavevector k_F associated with the free electrons among other things. We will first examine the distance dependence of the RKKY interlayer coupling between two ferromagnetic layers, F_1 and F_2.

To obtain the interlayer coupling energy between magnetizations in F_1 and F_2, it is necessary to consider the effect of all the spins in F_2 on a given spin in F_1. When a magnetic layer F_1 interacts with conduction electrons of the host (nonmagnetic) material, it introduces a polarization of spins which is propagated across the host into a neighboring magnetic layer F_2 and interacts with it. This is the origin of the "exchange coupling" between F_1 and F_2. This coupling energy per unit area, E/A, between the two ferromagnets is proportional to

$$K(\tilde{R}) = \int\limits_{\tilde{R}}^{\infty} dz \int\limits_{z}^{\infty} dx\, x^2 F(x)(1 - z/x), \tag{2.19}$$

where $\tilde{R} = 2k_F R$. The integral can be evaluated in closed form; the limiting form that is relevant to our discussion may be expressed as [26],

$$K(\tilde{R}) = \frac{\sin \tilde{R}}{\tilde{R}^2}\left(1 - \frac{27}{\tilde{R}^2}\right) - 5\frac{\cos \tilde{R}}{\tilde{R}^3} - O\left(\frac{1}{\tilde{R}^5}\right); \quad \tilde{R} \gg 1. \tag{2.20}$$

Note that for large distances R separating the slabs, the interaction energy decays as $1/R^2$ with a period given by π/k_F. As shown in Figure 2.7, this period is only a few Å, compared to nanometers observed in the oscillation periods of multilayer systems.

A theory similar in nature which incorporates the actual band structure was used to explain exchange coupling in real metallic multilayers. The crucial aspects of such a theory depends on the actual Fermi surface, its extrema and their distance from the Brullouin zone boundary associated with it. For realistic materials, their discrete lattice and band structures have to be taken into account and the above integrals be carried out using a primitive cell in reciprocal space. It should be noted here that early on, there was some skepticism in using such approaches. In one of the earliest discussions of the RKKY theory as applied to metallic multilayer systems, Bruno and Chappert [28] gave a fairly

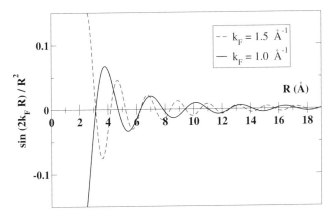

Figure 2.7.   Oscillations in exchange coupling energy between two ferromagnetic slabs obtained from Equation (2.20). Note that the period of oscillations in the interaction energy is only a few Å here. Compare with Figure 1.14 in Chapter 1 where the observed period (in Co/Cu) is of the order of a nanometer.

Table 2.2   Comparison of (long and short) oscillation periods measured using either SMOKE or SEMPA with those expected from the critical spanning vectors extracted from Fermi surface measurements using the de Hass–van Alphen (dHvA) effect

Interface	Long period (ML)	Short period (ML)	Technique	Reference
Ag/Fe(100)	5.58	2.38	dHvA	[27]
	5.73 ± 0.05	2.37 ± 0.07	SEMPA	[30]
Au/Fe(001)	8.6 ± 0.3	2.48 ± 0.05	SEMPA	[31]
Cu/Co(100)	5.88	2.56	dHvA	[28]
	8.0 ± 0.5	2.60 ± 0.05	MOKE	[32]
	6.0 to 6.17	2.58 to 2.77	SEMPA	[33]
Cr/Fe(100)	11.1		dHvA	[34,35]
	12 ± 1		SEMPA	[36]
	12.5		MOKE	[37]

Some of these experimental methods are discussed in Chapter 5.

detailed description of how and why RKKY type theories, derived for free electrons, have to be modified.

There is a clear resemblance with the simplest form of RKKY interactions encountered between magnetic impurities. When this simplest form is applied, making a free electron approximation and assuming a uniform continuous spin distribution within the ferromagnetic layers as shown above, the RKKY theory predicts a single period of about one monolayer (ML). This is much shorter than the observed periods in the multilayer systems with oscillatory coupling as evident from Table 2.2. Bruno and Chappert pointed out that the discrete size of the spacer has to be taken into account (the so-called "aliasing" effect) and the coupling can be related to the topological features of the Fermi surface of the spacer. They made quantitative predictions for multilayer systems with spacers consisting of noble metals.

The conclusions from Edwards *et al.* [29] are also quite illuminating, although their theory is based on a one band tight-binding model for the spacer layers. The exchange coupling was obtained by calculating the energy difference between parallel and antiparallel magnetizations of two infinitely thick magnetic layers separated by a nonmagnetic spacer consisting of $N - 1$ atomic planes. The resemblance to de Hass–van Alphen effect is striking, as pointed out in this work, where 2-dimensional quantization perpendicular to an applied magnetic field is utilized. Here in the multilayer sandwich, the quantization is one-dimensional, in the direction perpendicular to the layering.

1. Period of oscillations in exchange coupling: this is determined by the factor, $\exp[2i\,Nak_z^0(\mu)]$. Here $k_z^0(\mu)$ is an extremal radius of the Fermi surface in the direction perpendicular to the layering, and the derivatives that contribute to the oscillations are evaluated at $k_z^0 = k_z^0(\mu, k_x^0(\mu), k_y^0(\mu))$. Due to the discrete thickness Na of the spacer, $k_z^0(\mu)$ can be replaced by $k_z^0(\mu) - \pi/a$, providing a way to obtain a long period when the Fermi surface is close to the zone boundary (i.e., when $k_z^0(\mu) - \pi/a$ is small).
2. The amplitude of the oscillations contains a factor arising from the curvature of the Fermi surface at its relevant extremal points.

3. The temperature dependence of the oscillations is determined by the velocities of the carriers at these extremal points.
4. The asymptotic decay at zero temperature decays as $1/N^2$.

The RKKY interaction is usually written as,

$$H_{ij} = J(\mathbf{R}_{ij})\mathbf{S}_i \cdot \mathbf{S}_j \tag{2.21}$$

where the exchange integral

$$J(\mathbf{R}_{ij}) = C \int d^3q \chi(\mathbf{q}) \exp(i\mathbf{q} \cdot \mathbf{R}_{ij}). \tag{2.22}$$

This coupling constant is really the Fourier transform of the electronic susceptibility, $\chi(\mathbf{q})$, of the host material where

$$\chi(\mathbf{q}) = V_0/(2\pi)^3 \sum_{n,n'} \int d^3k \frac{f(\epsilon_{n,k}) - f(\epsilon_{n',\mathbf{k+q+G}})}{-\epsilon_{n,\mathbf{k}} + \epsilon_{n',\mathbf{k+q+G}}}. \tag{2.23}$$

Note that the band structure enters the above equations through the eigenvalues ϵ. The smooth parts of the function $\chi(\mathbf{q}_\parallel, q_z)$ give rise to short range oscillations, while the singular parts (in q_z) produce long range oscillations in the z direction. This was the key to understanding the long periods of oscillations and provided the answer to the initial concerns (or doubts) raised with regard to an interpretation based on a RKKY type model.

In Fourier analysis, this effect is referred to as "aliasing", which can be described as follows [28]. Let $f(z)$ be a function of a continuous variable z and $F(q)$ be its Fourier transform; suppose f_N is a series obtained by sampling f at equally spaced, discrete points $z_N = Nd$ with N being an integer. The Fourier representation $\tilde{F}(q)$ of f_N is a periodic function of q with period $2\pi/d$. Hence the values of q in the interval $[-\pi/d, \pi/d]$ are sufficient to describe f_N. Thus, there is an effective period Λ associated with a wavevector q, given by

$$2\pi/\Lambda = \left| q - m\frac{2\pi}{d} \right| \tag{2.24}$$

for suitable integers m, such that the q vector is folded into the above interval (see Figure 2.8). This effective period can be larger than the oscillatory periods expected from the RKKY theory. Also, the discrete (lattice) spin distribution, instead of a continuous one, gives rise to a Fermi surface in the extended zone scheme and thereby wavevectors (Figure 2.9) k^μ_z and k^ν_z can contribute the oscillations when $q_z = k^\mu_z - k^\nu_z$ satisfies a certain criterion, usually referred to as either "complete or partial nesting".

For more information on interlayer exchange coupling, there are a number of good review articles. The series Ultrathin Magnetic Structures I–IV consists of review articles on the general topic of magnetic multilayers. Chapter 2 of Ultrathin Magnetic Structures II contains four review articles written around 1993 covering various aspects of interlayer exchange coupling [38]. Volume III contains yet another review article, written in 2002. Volume 200 of the Journal of Magnetism and Magnetic Materials consists of a series of review articles covering much of magnetism and includes many related to magnetic multilayers. The article on interlayer exchange coupling [39] focuses on the comparison between theory and experiment. Other general review articles on interlayer exchange coupling include Refs. [40–42]. For a collection of theoretical and experimental results for specific systems, see [48]. The system Fe/Cr is sufficiently rich that it has generated

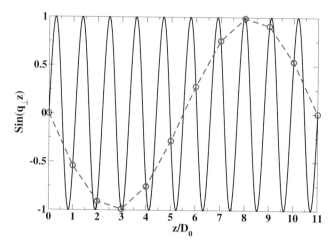

Figure 2.8.   Aliasing explained: the wave vector $q_\perp = 2\pi/(1.1d)$ associated with the periodic function can be folded into the interval $[-\pi/d, \pi/d]$ by subtracting an integral multiple of $2\pi/d$ where d is the spacing of periodically sampled points in real space, i.e., $2\pi/(1.1d) - 2\pi/d = -2\pi/11d$ giving rise to a long period $(11d)$ of oscillation in real space, as evident from the figure where $D_0 = d$.

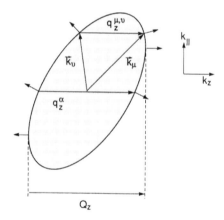

Figure 2.9.   The relevant **k** vectors related to the Fermi surface discussed in Ref. [28], reproduced with the permission of the American Physical Society.

review articles on its own [43,44]. Most of the articles above focus on transition metal multilayers, for reviews of rare earth multilayers, see [46,47]. Photoemission studies of quantum well states in magnetic multilayers are discussed in detail in Refs. [49,45]. Bi-quadratic coupling is reviewed in Ref. [50].

2.7.   Local spin-density functionals

Local spin-density functionals have been in use for quite some time now in calculations of band structure related properties of itinerant materials. A formal introduction to the Density Functional Theory, based on two theorems proven by Hohenberg and Kohn, is given in Chapter 6 of this volume. Nowadays most electronic structure work on weakly correlated systems is based on various (local or gradient corrected) approximations to density functionals. For a Hamiltonian operator H of a many Fermion system, the expectation value,

$$E\{\psi\} = \langle \psi | H | \psi \rangle \geqslant E\{\psi_0\} = E_0$$

represents the energy in the many-electron state $\psi = \psi(\mathbf{r}_1, \mathbf{r}_2, \ldots, \mathbf{r}_N)$. According to Levy [52], a functional, $\bar{E}\{n\}$, defined as,

$$\bar{E}\{n\} = \min_{\psi} \langle \psi | H | \psi \rangle \tag{2.25}$$

where the minimization is carried out with respect to the set $\{\psi\}$ whose members reproduce the given charge density,

$$n(\mathbf{r}) = \langle \psi | \sum_i \delta(\mathbf{r} - \mathbf{r}_i) | \psi \rangle. \tag{2.26}$$

Von Barth and Hedin [21] showed how to generalize the DFT to itinerant spin systems, by introducing spin-polarization through a magnetization density

$$\mathbf{m}(\mathbf{r}) = \langle \psi | \sum_i \sigma_i \delta(\mathbf{r} - \mathbf{r}_i) | \psi \rangle. \tag{2.27}$$

The operators σ_i have the well-known matrix representation, in terms of the Pauli spin matrices. If the direction of the magnetic moment can be fixed by an applied magnetic field along (say) the $+z$ direction, then spin densities n_+ for majority and n_- for minority spins can be expressed as,

$$n_\pm(\mathbf{r}) = \langle \psi | \sum_i \frac{1}{2}(\mathbf{1} \pm \sigma_{zi}) \delta(\mathbf{r} - \mathbf{r}_i) | \psi \rangle, \tag{2.28}$$

with

$$n(\mathbf{r}) = n_+(\mathbf{r}) + n_-(\mathbf{r}) \quad \text{(i.e., charge density)} \tag{2.29}$$

and

$$m(\mathbf{r}) = n_+(\mathbf{r}) - n_-(\mathbf{r}) \quad \text{(i.e., magnetization density)} \tag{2.30}$$

with spin moments pointing along the $\pm z$ directions, depending on whether $m(\mathbf{r})$ is positive or negative respectively at a given position $\mathbf{r}$.

At this point, one should realize that there has to be a local potential that breaks the degeneracy of the majority and minority spins. This is, of course, arising from the spin-density functionals, discussed in the literature, such as the already mentioned von Barth–Hedin functional [21]; i.e., in the density functional energy expression, there has to be a potential term

$$V_{xc,\sigma} = \frac{\delta E_{xc}\{n_+, n_-\}}{\delta n_\sigma(\mathbf{r})} \tag{2.31}$$

where

$$E_{xc} = \int \mathbf{dr}\, n(\mathbf{r}) \epsilon_{xc}(n_+, n_-) \tag{2.32}$$

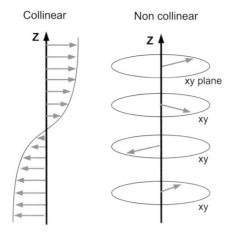

Figure 2.10.   Collinear vs. noncollinear magnetism.

with ϵ_{xc} representing a local exchange–correlation energy. The total energy within such a scheme,

$$E_{tot} = T_{n.i.}[n] + U[n] + E_{xc}[n_+, n_-],\qquad(2.33)$$

is a sum of the non-interacting kinetic term ($T_{n.i.}$), Hartree term (U, resulting from classical Coulomb interactions between charges) and the spin-dependent exchange–correlation energy, E_{xc}. The exchange part ϵ_x of ϵ_{xc}, expressed in the local density form, is proportional to spin-polarized densities as

$$\epsilon_x \propto \left(n_+^{4/3} + n_-^{4/3}\right)/n,\qquad(2.34)$$

which shows that the separate spin densities are necessary for spin polarized, local density calculations. Note here that these spin densities have to include the core electron densities (i.e., the total spin up and the total spin down densities are needed to evaluate the spin-split potentials).

The single particle equation,

$$\left\{\frac{1}{2}\nabla^2 + V_{Hartree} + V_{xc}\right\}\psi_{i\sigma} = \lambda_{i\sigma}\,\psi_{i\sigma},\qquad(2.35)$$

yields spin-dependent one particle energies $\lambda_{i\sigma}$ and wave functions $\psi_{i\sigma}$.

2.8.   Helical magnetic configurations: non-collinear magnetism

Magnetism in multilayer films is not limited to simple ferro- or antiferromagnetic alignments. In this section, we briefly discuss how to (theoretically) deal with somewhat more complex magnetic arrangements such as the noncollinear magnetic structures (Figure 2.10) encountered in multilayers from first principles. The spin dependent part of the effective one-particle potential, $V_{xc}(n_+, n_-)$, is tied to the Pauli spin matrix σ_z, implying that there is only one global set of axes for the spin. In a spin-wave or a spiral-like configuration, the local spin axes may be rotated with respect to such global spin axes. In most spin density functional calculations carried out in the past, this freedom is suppressed, so that only collinear magnetic configurations (Figure 2.10) are possible. However, this constraint can be removed using a rotation in spin space such as, $U_n^+ \sigma_z U_n$ with the spin rotation matrix U_n, in the nth layer given by

$$U_n = \begin{pmatrix} \cos(\theta_n/2)\exp(i\phi_n/2) & \sin(\theta_n/2)\exp(-i\phi_n/2) \\ -\sin(\theta_n/2)\exp(i\phi_n/2) & \cos(\theta_n/2)\exp(-i\phi_n/2) \end{pmatrix}. \tag{2.36}$$

Clearly, a larger unit cell is necessary to capture a spiral configuration, as shown in Figure 2.10. Note that the direction of the spin relative to the lattice is indeterminate; i.e., there still remains the question of the direction of the total moment with respect to crystallographic axes.

2.9. Orbital and multiplet effects

Unlike spin moments, orbital moments in solids are usually quenched (i.e., suppressed), due to the crystal field as well as the hybridization effects arising from the neighboring atoms. However, when the bulk crystal symmetry is broken, such as at an interface or a surface, enhanced orbital moments can be (and have been) separated from spin moments and detected, using experimental techniques such as magnetic X-ray circular dichroism (MXCD). For example, the ratios of orbital to spin moments, M_L/M_S, have been detected experimentally for Co/Cu(100) as a function of the Co layer thickness in Ref. [51] and these show an enhancement of the orbital effects at the top (surface) Co layer. From our discussion of MXCD in a later chapter, it should be clear that such an orbital contribution arises from an open shell of electrons where states with the orbital magnetic quantum numbers m and $-m$ are no longer degenerate. A strong spin–orbit interaction and a magnetic field which breaks the time reversal symmetry in the open shells will also yield enhanced orbital moments. In band structure calculations, the usual symmetry between $\mathbf{k} \rightarrow -\mathbf{k}$ is broken in this case and, it is imperative to include all such $\mathbf{k}$ vectors when calculating orbital moments.

However, there are other difficulties to overcome when calculations are carried out for orbital moments. For example, there are two-electron terms that contribute to spin–orbit energy [53]; orbital polarization and other (Hund's rule) related orbital effects are not rigorously dealt within a standard DFT treatment. The general consensus is that this is not a fundamental flaw of DFT. What is missing here is an orbital/spin current based DFT. This should be clear (see discussion below) since there is no information specific to orbital polarization in a standard density functional energy expression. One can recall that (even) for spin polarization, it was necessary to begin with a probe that can distinguish between different spin projections (as discussed in the chapter on effective action), or an energy expression with explicit spin polarization included. Since orbital effects are directly tied to orbital angular momentum, understanding orbital polarization in atoms would be quite helpful in this regard. To our knowledge, there are not that many rigorously justifiable, calculation schemes that address this issue, although current density functionals have been studied for some time now (see the last section of this chapter). The rest of this chapter will provide an examination of related issues with results from atomic multiplet calculations and some proposed solutions.

Hartree–Fock (HF) theory has had some success describing multiplets in atoms partly because the related exchange energy (from nonspherical terms) depends on whether the orbital moment of an electron is parallel or antiparallel to another electron [53]. Early on, Racah (see Ref. [54]) introduced linear combinations of the Slater F^k integrals to deal with intra-atomic electron interactions in atoms. He used labels (now called Racah coefficients) such as B, a combination of F^2 and F^4 appropriate for the d^n shell, or $E^{(3)}$, a combination of F^2, F^4 and F^6 appropriate for a f^n shell. The Slater integrals $F^k(l, l')$ are defined as

$$F^k(l, l') = e^2 \int\limits_{0}^{\infty}\int\limits_{0}^{\infty} \frac{r_<^k}{r_>^{k+1}} \phi_l^2(r_1)\phi_{l'}^2(r_2)\, dr_1\, dr_2 \tag{2.37}$$

for radial orbitals with angular momenta l and l'.

To obtain a better understanding of these HF exchange energy contributions, consider the d^2 configuration (in which Ti^{2+} is found) [53]. This configuration forms ^3F, ^3P, ^1S, ^1D, and ^1G multiplet levels. The first and the last multiplet levels here, have single determinant wave functions. Those are:

$$\Psi\left(^3F\right)_{31} = \Phi\left(2^+, 1^+\right), \tag{2.38}$$

$$\Psi\left(^3F\right)_{21} = \Phi\left(2^+, 0^+\right), \tag{2.39}$$

$$\Psi\left(^1G\right)_{40} = \Phi\left(2^+, 2^-\right). \tag{2.40}$$

The arguments above, for example in the determinant Φ in Equation (2.38), refer to $m_l = 2$ with an up spin (+) for the first electron and $m_l = 1$, also with an up spin (+) for the second electron. The HF exchange energy for these two electrons is given by [54]

$$E_{exch} = -\frac{6}{49}F^2(dd) - \frac{5}{441}F^4(dd), \tag{2.41}$$

where F^2 and F^4 are Slater radial integrals (defined in Equation (2.37)) which involve a common d radial function for the electrons of the open shell. Now, if the m_l value of the second electron is flipped from $+1$ to -1, with the new determinant belonging to a different multiplet, the exchange energy is,

$$E_{exch} = -0F^2(dd) - \frac{35}{441}F^4(dd). \tag{2.42}$$

This difference is essentially due to the nonspherical terms present in the two electron integrals arising from an expansion of $1/|\mathbf{r} - \mathbf{r}'|$ [54].

The electron charge and spin densities are the same for the two determinants but exchange energies are different (while the direct Coulomb energy stays the same). This example shows that for DFT based approaches (such as LDA or even GGA), where only charge and spin densities are used, orbital polarization (or current) terms need to be included if orbital moments are to be calculated at all.

The following is such an attempt, introduced in Ref. [55]. In terms of the Racah coefficients, B and $E^{(3)}$, energy splittings between states with *maximum spin but varying L* can be expressed. For such states, the energy of an individual level corresponds to

$$E_{av} + \chi L(L + 1)\zeta, \tag{2.43}$$

where E_{av} is the mean energy of the configuration, χ a simple constant $(-1/2)$ and $\zeta = B$ for the case of a d shell or $\zeta = E^{(3)}$ for an f shell atom. The above L dependence is clearly absent in an ordinary DFT-like treatment; such an orbital dependence will lead, from a variational treatment, to an orbital polarization term in an effective one particle treatment as follows. A term such as

$$\chi \langle L \rangle \zeta \hat{m}_l \qquad (2.44)$$

may be introduced for orbital calculations. There are schemes which introduce such a term to LSDA calculations. Unlike in a pure LSDA treatment, such terms together with the spin–orbit interaction leads to orbital polarization and related orbital, magnetic effects encountered in magnetic circular dichroism experiments, where $\pm m$ effects can be distinguished. Use of this term, in a band structure calculation with spherical potentials (as in [55]), avoids the nonspherical exchange issues discussed above expected to deal with orbital polarizations.

It is easy to see that the above polarization term is convenient and easy to implement. However, even for atoms, the applicability of expression (2.43) is rather limited as shown below [53].

1. Configurations d^2, d^3, d^7, d^8:
    The states of maximum spin are F and P levels whose separate energies do not satisfy the energy expression above, but their difference may be written as,

    $$E(F) - E(P) = \chi \Delta \left[L(L+1) \right] B,$$

    where $\chi = -3/2$, the coefficient used in transition metal oxide calculations.
    Among the states having less than maximum spin, those with maximum L are not the lowest in energy. Note that Hund's second rule does not apply to states with spin less than the maximum and no single value of the coefficient χ describes such splitting between states of lower S.
2. Configurations d^4, d^5, d^6, f^6, f^7, f^8:
    These have only a single multiplet state of maximum S. In the manifolds of lower spin, the states of maximum L often lie lowest. In such cases, χ is negative when estimating the energy with respect to the others. However, no single value of χ is sufficient to define to all the splittings between all the L states of any one of these manifolds.

Similar conclusions can be drawn for other configurations involving f electrons [53]. The lesson to be learned here is that, although it is very convenient to use, the expression involving a $L(L+1)$ contribution to the total energy of a multiplet may, more often than not, only turn out to be minimally justifiable from a theoretical point of view.

2.10. Current density functional theory

In the previous section, deficiencies in the available Kohn–Sham methods when trying to deal with orbital (current) effects were pointed out. Spin density functionals simply do not carry orbital (current) information and hence cannot be used to treat situations where both spins and currents are involved. Such situations are frequently encountered in magnetic materials and applications such as spin valves. If the spin–orbit effects are significant, it becomes imperative to correctly treat these when studying transport properties, among others. In order to treat currents within the DFT formalism, Vignale and Rasolt [56], introduced a current-spin density functional formalism during the late 1980s. Unfortunately, due to a lack of practical schemes based on a sufficiently simple approximation (such as the LDA to DFT), there has been limited progress towards actual calculations.

2.10.1.   Current density in exchange–correlation of atomic states

Here we follow the discussion of Becke [57] with regard to open shell atoms in the first row of the periodic table. As discussed earlier, a given atomic state may not be representable by a single determinant and multideterminantal states cannot be dealt within the standard DFT. Even in single determinantal cases, there are situations where degenerate orbitals of different (orbital components) m_l do not have the same electron densities. Consider one p electron in an open shell atom, such as Boron. The electron can occupy one of the three p orbitals, $m_l = -1, 0, +1$. The complex spherical harmonics associated with these orbitals produce two different densities for $m = 0$ and $m = \pm 1$ and approximate density functionals, such as the LSDAs or GGAs, will yield different exchange–correlations energies and hence will not reproduce exact degeneracies of the states. If we consider real spherical harmonics, the problem would be solved, but not for d and f orbitals. There are other instances where similar or related problems have been pointed out [53].

Now $m_l = \pm 1$ orbitals have nonzero current density and if this is taken into account, Becke [57] demonstrated that the exchange–correlation energy difference between $M_L = \pm 1$ and $M_L = 0$ states undergoes a dramatic reduction. For two-electron p^2 configurations, the ground state is ^3P. Treating both spins up (or down), the single determinants with $m_l = 0$ and $m_l = \pm 1$, and the single determinant with $m_l = 1$ and $m_l = -1$ are valid ^3P states. The former has $M_L = \pm 1$ with a nonzero current while the latter has $M_L = 0$ with zero current. When the current density functional proposed by Becke is implemented, the exchange–correlation energy difference between such states (which are supposed to be degenerate) is reduced below 1 kcal/mol (from GGA values that are several kcals/mol) for a number of small atoms in the first and second rows of the periodic table.

In the absence of magnetic fields and spin–orbit effects, the electronic ground state of oxygen is a ninefold degenerate ^3P state of valence configuration $2s^2, 2p^4$. Using previous notation (from Equation (2.38)) in a nonzero magnetic field, the $^3P_2$ state with total magnetic quantum number $M_J = -2$ [58]

$$\Psi\left(^3P\right)_{-1,-1} = \Phi\left(-1^-, 0^-, 1^-, -1^+\right), \tag{2.45}$$

is the ground state (single determinantal state with $p_{m_l m_s}$ orbitals as indicated in Equation (2.45)). A Kohn–Sham treatment of this state, even in the absence of a magnetic field, leads to a lifting of the degeneracy of the p orbitals because of the partially filled p shell accompanied by a cylindrical charge density.

All of the above arguments point out deficiencies in the present day utilizations of the DFT methods. An important issue is a practical implementation that can accurately calculate orbital and spin dependencies. Recently, an exact-exchange spin-current density functional theory has been proposed [58] to calculate spin currents. The Hamiltonian of an electronic system in a magnetic field is expressed as

$$\hat{H} = \hat{T} + \hat{V}_{ee} + \int d\mathbf{r} \, \Sigma^T \mathbf{V}(\mathbf{r}) \hat{\mathbf{J}}(\mathbf{r}) \tag{2.46}$$

with $\hat{T}$ and $\hat{V}_{ee}$ being the operators of the kinetic energy and the electron–electron repulsion. The 4-vector Σ has unit 2×2 matrix and Pauli spin matrices as its components while the vector $\hat{\mathbf{J}}(\mathbf{r})$ has the density operator and x, y, z components of the current operator as its four components. The 4×4 matrix $\mathbf{V}(\mathbf{r})$ is composed of matrix elements $V_{\mu\nu}$ containing the external potential v_{ext} and magnetic field $\mathbf{B}$ with its vector potential $\mathbf{A}$. A Hohenberg–Kohn functional is defined so that

$$F[\rho] = \min_{\Psi \to \rho} \langle \Psi | \hat{T} + \hat{V}_{ee} | \Psi \rangle, \tag{2.47}$$

where the minimization wavefunctions are constrained to not only to give the correct density but also to yield all the 16 components of the spin-current density.

For an oxygen atom in a magnetic field along (say) the z axis, this approach shows that there are orbital currents around this axis. It appears to correctly generate the spin and orbital angular momenta and align them along the magnetic field. Practical schemes that can deal with spin-current densities, similar to the one described above, will hopefully be available in the near future to treat more complex problems.

References

[1] N.W. Ashcroft, N.D. Mermin, Solid State Physics (Saunders Publishers, 1976).
[2] J. Korringa, Physica **13**, 392 (1947);
    W. Kohn, N. Rostoker, Phys. Rev. **94**, 1111 (1954).
[3] O.K. Andersen, Solid State Commun. **13**, 173 (1973).
[4] O. Jepsen, Phys. Rev. B **12**, 2988 (1975);
    O. Jepsen, O.K. Andersen, A.R. Mackintosh, Phys. Rev. B **12**, 3084 (1975).
[5] D.D. Koelling, Arbman, J. Phys. F **5**, 2041 (1975).
[6] O. Jepsen, J. Madsen, O.K. Andersen, Phys. Rev. B **18**, 605 (1978).
[7] H. Krakauer, M. Posternak, A.J. Freeman, Phys. Rev. B **19**, 1706 (1979).
[8] H.L. Skriver, The LMTO Method (Springer-Verlag, New York, 1984).
[9] G.W. Fernando, B.R. Cooper, M.V. Ramana, H. Krakauer, C.Q. Ma, Phys. Rev. Lett. **56**, 2299 (1986).
[10] G.W. Fernando, J.W. Davenport, R.E. Watson, M. Weinert, Phys. Rev. B **40**, 2757 (1989).
[11] A.J. Appelbaum, D.R. Hamann, Rev. Mod. Physics **48**, 479 (1976).
[12] O.K. Andersen, Phys. Rev. B **12**, 3060 (1975).
[13] T. Loucks, Augmented Plane Wave Method (Benjamin, New York, 1976).
[14] M. Weinert, J. Math. Phys. **22**, 2433 (1981).
[15] D.R. Hamann, L.R. Mattheiss, H.S. Greenside, Phys. Rev. B **24**, 6151 (1981).
[16] J.F. Cornwell, Group Theory and Electronic Energy Bands in Solids (North-Holland, Amsterdam, 1969).
[17] P. Hohenberg, W. Kohn, Phys. Rev. B **136**, 864 (1964).
[18] W. Kohn, L.J. Sham, Phys. Rev. A **140**, 1133 (1965).

[19] R.O. Jones, O. Gunnarsson, Rev. Mod. Phys. **61**, 689 (1989);
P. Fulde, Electron Correlations in Molecules and Solids (Springer-Verlag, New York, 1995);
K. Held, D. Vollhardt, European Phys. J. B **5**, 473 (1998);
I. Yang, S.Y. Savrasov, G. Kotliar, Phys. Rev. Lett. **87**, 216405 (2001).

[20] U. von Barth, L. Hedin, J. Phys. C **5**, 1629 (1972).

[21] U. von Barth, L. Hedin, Phys. Rev. B **4**, 1629 (1972);
O. Gunnarsson, B.I. Lundqvist, Phys. Rev. B **13**, 4274 (1976).

[22] S. Ohnishi, A.J. Freeman, M. Weinert, Phys. Rev. B **28**, 6741 (1983).

[23] M. Weinert, S. Blügel, in: L.H. Bennet, R.E. Watson (Eds.), Magnetic Multilayers, World Scientific (1994).

[24] V.L. Moruzzi, J.F. Janak, A.R. Williams, Calculated Electronic Properties of Metals (Pergamon, New York, 1978).

[25] M.A. Ruderman, C. Kittel, Phys. Rev. **96**, 99 (1954);
T. Kasuya, Prog. Theor. Phys. **16**, 45 (1956), 58;
K. Yosida, Phys. Rev. **106**, 893 (1957).

[26] W. Baltensperger, J.S. Helman, Appl. Phys. Lett. **57**, 2954 (1990).

[27] P. Bruno, C. Chappert, Phys. Rev. Lett. **67**, 1602 (1991).

[28] P. Bruno, C. Chappert, Phys. Rev. B **46**, 261 (1992).

[29] D.M. Edwards, J. Mathon, R.B. Muniz, M.S. Phan, Phys. Rev. Lett. **67**, 493 (1991).

[30] J. Unguris, R.J. Celotta, D.T. Pierce, J. Magn. Magn. Mater. **127**, 205 (1993).

[31] J. Unguris, R.J. Celotta, D.T. Pierce, J. Appl. Phys. **75**, 6437 (1994).

[32] M.T. Johnson, S.T. Purcell, N.W.E. McGee, R. Coehoorn, J. aan de Stegge, W. Hoving, Phys. Rev. Lett. **68**, 2688 (1992).

[33] W. Weber, R. Allenpach, A. Bischof, Europhys. Lett. **31**, 491 (1995).

[34] M.D. Stiles, Phys. Rev. B **54**, 14679 (1996).

[35] L. Tsetseris, B. Lee, Y.-C. Chang, Phys. Rev. B **55**, 11586 (1997);
L. Tsetseris, B. Lee, Y.-C. Chang, Phys. Rev. B **56**, R11392 (1997).

[36] J. Unguris, R.J. Celotta, D.T. Pierce, Phys. Rev. Lett. **67**, 140 (1991).

[37] E.E. Fullerton, M.J. Conover, J.E. Mattson, C.H. Sowers, S.D. Bader, Phys. Rev. B **48**, 15755 (1993).

[38] K.B. Hathaway, A. Fert, P. Bruno, D.T. Pierce, J. Unguris, R.J. Celotta, S.S.P. Parkin, in: B. Heinrich, J.A.C. Bland (Eds.), Ultrathin Magnetic Structures, vol. II, Springer-Verlag, Berlin, p. 45 (1994), Chapter 2.

[39] M.D. Stiles, J. Magn. Magn. Mater. **200**, 322 (1999).

[40] J.C. Slonczewski, J. Magn. Magn. Mater. **150**, 13 (1995).

[41] Y. Yafet, in: L.H. Bennett, R.E. Watson (Eds.), Magnetic Multilayers, World Scientific, Singapore (1994).

[42] P. Bruno, J. Phys. Condens. Matter **11**, 9403 (1999).

[43] D.T. Pierce, J. Unguris, R.J. Celotta, M.D. Stiles, J. Magn. Magn. Mater. **200**, 290 (1999).

[44] R.S. Fishman, J. Phys. Condens. Matter **13**, R235 (2001).

[45] Z.Q. Qiu, N.V. Smith, J. Phys. Condens. Matter **14**, R169 (2002).

[46] C.F. Majkrzak, J. Kwo, M. Hong, Y. Yafet, D. Gibbs, C.L. Chein, J. Bohr, Adv. Phys. **40**, 99 (1991).

[47] J.J. Rhyne, R.W. Erwin, in: K.H.J. Buschow (Ed.), Magnetic Materials, vol. 8, North-Holland, Elsevier, Amsterdam, p. 1 (1995).

[48] D.E. Burgler, P. Grunberg, S.O. Demokritov, M.T. Johnson, in: K.H.J. Buschow (Ed.), Handbook of Magnetic Materials, vol. 13, Elsevier, Amsterdam (2001).

[49] F.J. Himpsel, Phys. Rev. B **44**, 5966 (1991);
J.E. Ortega, F.J. Himpsel, Phys. Rev. Lett. **69**, 844 (1992);
J.E. Ortega, F.J. Himpsel, G.J. Mankey, R.F. Willis, Phys. Rev. B **47**, 1540 (1993).

[50] S.O. Demokritov, J. Phys. D **31**, 925 (1998).

[51] M. Tischer, et al., Phys. Rev. Lett. **75**, 1602 (1995).

[52] M. Levy, Phys. Rev. A **26**, 1200 (1982).

[53] M. Weinert, R.E. Watson, G.W. Fernando, Phys. Rev. A **66**, 032508 (2002).

[54] E.U. Condon, J.H. Shortley, The of Atomic Spectra (Cambridge University Press, London, 1953); in particular, see Table 1^6 and Chapter 8.

[55] M.S. Brooks, P.J. Kelly, Phys. Rev. Lett. **51**, 1708 (1983);
O. Eriksson, B. Johansson, M.S. Brooks, J. Phys.: Condens. Matter **1**, 4005 (1989).

[56] G. Vignale, M. Rasolt, Phys. Rev. Lett. **59**, 2360 (1987);
G. Vignale, M. Rasolt, Phys. Rev. B **37**, 10685 (1988).

[57] A.D. Becke, J. Chem. Phys. **117**, 6935 (2002).

[58] S. Rohra, A. Görling, Phys. Rev. Lett. **97**, 013005 (2006).

Chapter 3

Thin Epitaxial Films: Insights from Theory and Experiment

3.1. Metastability and pseudomorphic growth

Metastability is encountered in the materials world more often than one would naively assume. From a theoretical point of view, a suitable free energy, when monitored as a function of various parameters such as lattice parameters, temperature or pressure, is a complicated function of many variables with possibly numerous local minima. Metastability that is discussed in this chapter has to do with the existence of such local minima. In order to crossover from one local minimum to a neighboring one, the system needs to overcome an energy barrier. Such barriers are usually referred to as activation barriers. Hence extra energy (in the form of thermal or other) is necessary for the system to sample various regions of the free energy surface. Growth conditions clearly play a vital role in determining the resulting metastable phase.

Transition metals provide a simple, yet rich, platform for studying metastable phases. In the experimental world, such phases can be achieved through various means. Alloying is one popular way to pursue metastable phases, although it is not the cleanest, due to the presence of some unwanted element(s). The idea is to extrapolate back to the pure system using the results from the (dilute) alloyed compound. Another method is to form precipitates in a host matrix, which could play a pivotal role in the final structure of the precipitate. Compared to the above two, epitaxial growth of a material on a well defined substrate provides a much more manageable as well as clean system for characterizing and studying its physical properties. A layer by layer, pseudomorphic growth (i.e., in registry with the substrate) usually provides insight into numerous size, surface and interface dependent properties of metastable structures.

In previous chapters, we have discussed magnetic multilayer films that are useful in read/write devices. In such applications, a clean, pseudomorphic growth is highly desirable, due to the sensitivity of magnetic interactions to structural and other defects. On the other hand, for heterogeneous catalysis, rough films with island formation would be considered desirable. Hence surface morphology plays an essential role when applications are being considered while the growth mode, for the most part, determines morphology. There are several standard (phenomenological) growth modes that are usually defined and frequently encountered. These are:

1. Frank–van der Mere growth [1],
2. Volmer–Weber growth [2],

3. Stranski–Krastanov growth [3].

The Frank–van der Mere growth mode is a *layer-by-layer* growth, i.e., a given layer has to be completed in order to begin the next, adjacent layer. Volmer–Weber process is defined by the formation of three dimensional islands on the surface, while the Stranski–Krastanov growth may be identified as a combination of the other two modes, i.e., a layer-by-layer growth accompanied by island formation. Note that surface defects, such as steps, as well as ambient conditions can also have a significant impact on the overlayer growth on a substrate. At high enough temperatures, adatoms can move rather rapidly and easily sample various pre-existing defects on a substrate and possibly get trapped. Such a growth process is likely to avoid the formation of islands, at least at the initial stages. Also, at such temperatures there could be significant diffusion into the substrate. Due to these reasons, elevated temperatures are not suitable for ordered growth processes. Here we do not intend to describe these growth processes in detail and the interested reader is referred to Ref. [4] and the above listed references on growth modes.

3.2. A brief introduction to kinetics

It is essential to realize that important stages of aforementioned growth modes are driven or controlled by kinetic processes, i.e., nonequilibrium phenomena and not simply equilibrium thermodynamics. Hence, as already mentioned, temperature as well as other ambient conditions at the surface usually play a significant role during growth. The equilibrium state is reached only after the kinetic steps, such as migration, diffusion, nucleation, evaporation, follow a certain path determined by various energy barriers and external conditions. This equilibrium state can easily be a metastable one, indicating that the kinetic processes could not access other states due to high energy barriers and ambient conditions. In brief, kinetic processes do not necessarily select the most stable, lowest energy state but the most easily (kinetically) accessible one.

As one would naively expect, diffusion of (impurity) atoms on a surface is different from bulk diffusion. The activation barriers E_s at the surface are much smaller that those found in the bulk and are related to the diffusion coefficient D, given by,

$$D = (\text{const.}) \exp(-E_s / k_B T). \tag{3.1}$$

Bulk diffusion usually happens through the motion of vacancies and interstitial atoms. Adatoms on a surface diffuse according to a process similar to a random walk, although even here diffusion can be mediated by defects such as vacancies. The migration of these species on a surface will depend on the surface morphology and ambient conditions such as temperature among other things. For example, if the surface has steps, along such a step on a given surface, adatoms can undergo "edge diffusion". Here the path of diffusion depends crucially on whether the adatom encounters a step or not. Alternatively, migrating atoms on a surface can find another such atom and start to form islands. From the above, one can see that surface diffusion and migration of atoms can play complicated roles in the kinetics of growth. Our discussion in this regard is a very limited one and the interested reader is referred to various articles on related kinetic processes (see, for example, Refs. [5–7]).

3.3. Strain in hetero-epitaxial growth

Thin epitaxial films are known to exhibit a wide variety of fascinating fundamental properties due to the interplay of electronic structure, morphology, strain and magnetism [8]. Naturally, to better understand the physical phenomena at hand and build useful theories, idealized, simple systems are highly desirable. Let us consider two different elemental metals A and B. When the elemental metal A is epitaxially grown on the elemental metal substrate B, the crystallographic orientations of the surface B plays a central role in determining the stability and growth of the metallic film A. If there is a close match between the substrate crystallographic orientation with that of metal A, the epitaxial growth process is expected to be at least stable up to a few monolayers (ML). A strain related parameter, defined as the misfit

$$\alpha_{A/B} = \frac{a(A) - a(B)}{a(A)} \tag{3.2}$$

will be useful in characterizing the epitaxial growth. The misfit α is a simple way to identify the difference in the corresponding bulk lattice parameters; i.e., it refers to unstrained lattice mismatch between A and B. For example, $\alpha_{Cu/Ni} = (a(Cu) - a(Ni))/a(Ni) = 2.6\%$ while $\alpha_{W/Fe} = (a(W) - a(Fe))/a(Fe) = 10.4\%$. Depending on the magnitude of the mismatch, several growth scenarios may be observed. Clearly, if the ground state crystal structures of A and B are the same with close lattice matching, epitaxial growth could be sustainable up to tens of ML. If the lattice mismatch is neither very small nor large, then several layers of the deposit metal A can be grown with some strain present in the (newly grown) layers. In this case, there will be some critical thickness up to which the system can sustain epitaxial growth. For Ni/Cu, clean, pseudomorphic growth is found up to 10–11 ML [9] while in Fe/W, dislocations are formed within the first few layers of growth [10]. When the strain energy is significant, the film will try to relieve its strain through induced defect formation (see Figure 3.1). If the misfit is too large, it is highly unlikely that there will be any pseudomorphic growth. Instead, there will be misfit dislocations from the beginning of the growth process. In this case there will be no strain in the deposit.

There are various growth mechanisms that have been observed in metal on metal systems. A great deal of attention has been paid to Fe films grown on various metallic substrates, such as Fe/Cu, Fe/W and Fe/Mo. One major reason for this interest is the magnetism associated with Fe films, which may depend on the growth mechanism. One such mechanism is the two-dimensional island nucleation followed by coalescence as observed in Fe/Mo(110) and Fe/W(110). The temperature of the system plays a significant role during the growth process. Fe films grown on W(110) begin as an array of unconnected islands at 300 K [10,12]. At coverages higher than 0.6 ML, ferromagnetic order is observed by magnetic percolation through island coalescence and the first Fe layer has been observed to become ferromagnetic below 225 K with an in-plane easy axis of magnetization along the [1$\bar{1}$1] direction. A second layer island growth, free of defects, is also observed up to about 1.5 ML with an easy axis of magnetization out-of-plane and ferromagnetic at 300 K. Above 1.5 ML, misfit dislocations begin to form reducing the strain with an in-plane easy axis of magnetization. In Figures 3.2 and 3.3, STM images of island formation and dislocation lines during the growth of Fe films on Mo(110) are shown [11].

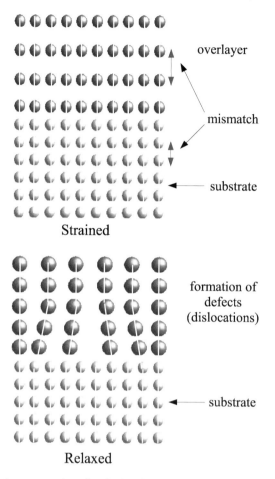

Figure 3.1.   A schematic representation of strained and relaxed epitaxial film growth on a given substrate. When the lattice mismatch is large, the film will try to relax and reduce its strain by creating various defects from the very first layer onwards.

3.4.   Alloy phase diagrams and defects

Binary (and other) alloy phase diagrams provide valuable insight into the initial steps of growth of a heteroepitaxial system. A binary alloy phase diagram of metals A and B that shows a tendency for strong compound formation is very likely to have a strong effect on the above growth. This is simply due to the fact that it will be energetically favorable for the atoms to inter-diffuse and gain energy by forming chemical bonds with different atoms instead of undergoing epitaxial growth. On the other hand, if the phase diagram shows no tendency for strong compound formation with a high degree of immiscibility, there will be very little inter-diffusion at the interface, if any.

Another problem that one is faced with when studying alloy phase diagrams as well as epitaxial growth is the presence of various defects. The statistical mechanics of an ordered, homogeneous thermodynamically stable compound is complicated by the fact

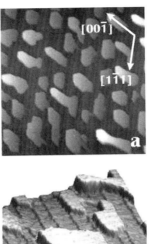

Figure 3.2. For Fe/Mo(111) (a) 5000×5000 Å^2 image of a 2.4 ML film grown on Mo(110) at 515 ± 15 K. (b) A three-dimensional representation of Fe islands showing the wedge shape formed as they propagate over several terraces while maintaining a flat (110) surface. Reproduced with the permission of the American Physical Society [11].

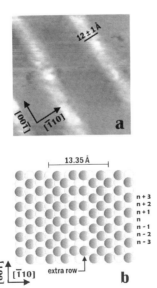

Figure 3.3. (a) 90×90 Å^2 STM image showing dislocation lines in the second layer of a 1.8 ML Fe film deposited at 340 ± 15 K. (b) A model of the dislocation due to the insertion of an extra row of Fe atoms along the [00$\bar{1}$] direction. The first layer Fe atoms are dark gray, while the second layer Fe atoms are light gray. Reproduced with the permission of the American Physical Society [11].

that various types of atomic defects must coexist in order to ensure that stoichiometry and homogeneity of the material are maintained. In epitaxial growth, creation of defects naturally leads to a reduction in strain.

Understandably, such theoretical studies of defect formation in ordered systems are nontrivial. If we assume that the system under consideration contains a low concentration of defects, then a noninteracting statistical mechanical model can be developed that describes a system in thermodynamic equilibrium with its defects. By minimizing an ansatz for the Gibbs free energy with respect to the number of defects, one can obtain expressions for the defect concentration within a first principles framework [14]. These expressions contain parameters ε_i^ν and υ_i^ν that represent the difference in energy and volume, respectively, between a cell containing a defect and an ideal cell. Since the volumes have been relaxed ($P = 0$) in the first-principles calculations, all the $P\upsilon_i^\nu$ terms in the free energy drop out. Furthermore, the energetics furnished by the first-principles calculations are used to obtain the ε_i^ν parameters. This is an acceptable approximation as long as the cells

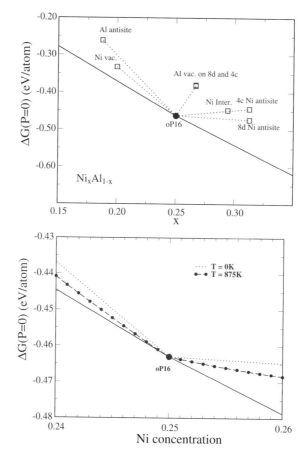

Figure 3.4.   Calculated energetics of point defects in the ordered compound NiAl$_3$. The top figure shows $T = 0$ values while the bottom figure shows that the defects shown cannot co-exist; i.e., NiAl$_3$ remains a strong line compound in agreement with experiment, published in Ref. [14].

used in the first principles calculation are large enough for defect-defect interactions to be small. Once these parameters have been calculated, the equations for the defect concentrations along with other constraints imposed on the system are solved for the chemical potentials of the constituent atoms and then the formation energies of the thermally activated defects can be found. In Figure 3.4, results from such calculations are shown. The defect structures examined here do not appear to be able to co-exist with the ordered alloy $NiAl_3$, and it remains a line compound (with no width around it in the phase diagram) in agreement with experiment.

3.5. Metastability and Bain distortions

In order to search for metastable structures from theory, one should be able to compute free energies of various configurations of atoms relevant to those of interest, and compare them. For ground state total energy calculations of various structures, there is now a well established (mean-field) method, namely the DFT-based band structure calculations. We will discuss a few simple examples of such calculations carried out for nonmagnetic as well as magnetic cases. It should be understood that these represent only a very small sample of what has been reported up to date. Theoretically, DFT based local spin density approximation (LSDA) work has been carried out in order to search for metastable (magnetic) structures. Morruzi and Marcus (Ref. [19]) and many others have reported such searches where the magnetic moment and unit cell volume are constrained when total energy is minimized. This will lead to a set of possible metastable structures and corresponding binding energy surfaces that are in the neighborhood of the global minimum of total energy. An illustrative example is provided by the binding energy surfaces of fcc and bcc Fe, as shown in Figure 3.11. Most of the early calculations were restricted to high symmetry, bcc, fcc (cubic) and hcp (hexagonal) structures (see the book by Moruzzi *et al.* [19]). However, it was realized that other lower symmetries, such as tetragonal structures, could also be quite relevant with regard to metastable phases grown on high index crystal surfaces.

   In 1924, Bain [20] introduced the concept of a tetragonal lattice distortion of a fcc lattice into a bcc lattice (and its reverse) in the context of the formation of Martensite in steel. This transformation, now referred to as the "Bain distortion", has been increasingly utilized to understand lattice distortions in cubic, tetragonal and other crystals, especially after the availability of reliable, first principles total energy calculations. One important reason for this resurgence is the ability of such calculations to probe paths that are experimentally inaccessible. Examining such paths can provide useful insights into experimentally accessible phases. For example, certain monatomic crystals which have a fcc (bcc) ground state might have an energy maximum along the Bain path at their corresponding bcc (fcc) phase (see Figure 3.5). In particular, Mehl and Boyer [21] have shown that the metals Al and Ir, which have fcc ground states, are elastically unstable in the bcc phase, i.e., the total energy reaches a maximum at the bcc point along the Bain path. There are many similar calculations that have been reported in the literature. (For example, Ref. [25] reports that the fcc elements Ni and Rh are unstable along the Bain path at the bcc point while the bcc elements Co, Cr, Mo, Nb, Ta and V are unstable along the Bain path at the fcc point. Also, Jona and Marcus [23,24] have reported that the fcc elements Cu and Pd are elastically unstable at their bcc point along the Bain path.)

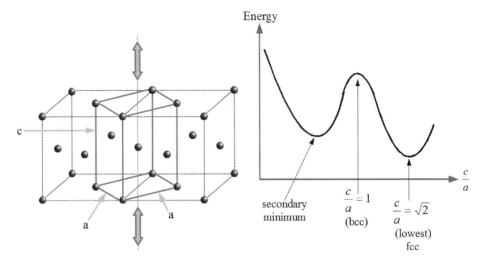

Figure 3.5.   A schematic illustration of the Bain distortion [20] and possible energy changes along the distortion path c/a in monatomic crystals. Tetragonal lattice parameters c and a are shown on the left. In the energy plot, note the presence of a secondary minimum away from the fcc and bcc extrema.

The existence of a minimum and a maximum implies that there has to be another minimum (see Figure 3.5) outside the classical fcc-bcc path of the Bain distortion [25]. Mehl *et al.* (Ref. [22]) argue that this secondary local minimum has to be unstable, except in a rare situation. However, experimentally it has been demonstrated that some of these unstable phases, such as body centered tetragonal structures, can be epitaxially grown on a suitable substrate. This clearly underscores the importance of the substrate for such systems, since free standing layers cannot be thermodynamically and elastically stable at low temperature.

Here we will briefly describe the study of local stability of nonequilibrium phases carried out by Craievich *et al.* [25], where the relative structural stability of transition metals and their alloys with respect to tetragonal distortions was investigated using DFT-based total energy calculations. According to this study, "nonequilibrium cubic structures" were found to be locally unstable with respect to such distortions, i.e., the cubic elastic constants, $c_{11} - c_{12}$, were found to be negative. Figure 3.6 shows the energy changes along the Bain path for the $3d$, $4d$, and $5d$ transition metal rows while Figure 3.7 is a similar plot for Ti and RuNb. A common feature of all these energy changes is that in addition to the minimum at the low energy structure, there exists a maximum for a (different) high symmetry structure along the Bain path.

In the above study, total energies of various transition metals in hypothetical tetragonal structures were studied along the volume conserving, Bain distortion in c/a, where c and a are standard, tetragonal lattice parameters. When the face centered cubic (fcc) structure is viewed as a special case of a body tetragonal cell, shown in Figure 3.5, a straightforward connection between the fcc and the bcc structures can be seen, as discussed in elementary texts. This well known Bain distortion takes a hypothetical crystal from a bcc ($c/a = 1$) to a fcc structure ($c/a = \sqrt{2}$).

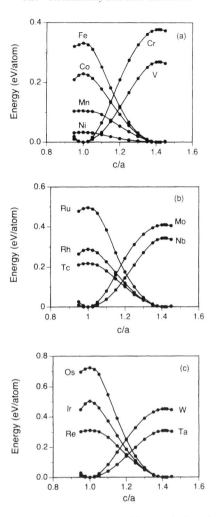

Figure 3.6. Self-consistent LMTO-ASA results for the transition metals along the tetragonal deformation path. Energies are relative to the low energy structure of the elements along the path for the (a) $3d$, (b) $4d$, and (c) $5d$ rows. Reproduced with the permission of the American Physical Society from Ref. [25].

If the volume is conserved along the path of this distortion, then, at the above (bcc and fcc) end points, the total energy must be an extremum under the conditions imposed. This can be easily seen, if we express the total energy of an orthorhombic system as $U(a, b, c)$ where a, b and c are the lattice constants. Then the total differential is

$$\delta U = \frac{\partial U}{\partial a}\delta a + \frac{\partial U}{\partial b}\delta b + \frac{\partial U}{\partial c}\delta c = 2\frac{\partial U}{\partial a}\delta a + \frac{\partial U}{\partial c}\delta c, \tag{3.3}$$

for the tetragonal structure where $a = b$. The conservation of volume implies that

$$a\delta c = -2c\delta a \tag{3.4}$$

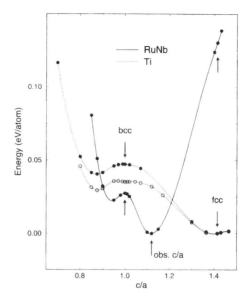

Figure 3.7.   Paramagnetic self-consistent full-potential LASTO results along the tetragonal Bain distortion at fixed volume. The RuNb calculations were done at the observed volume. For Ti, the results corresponding to both the observed volume (filled circles) and the calculated bcc volume (open circles) are shown. The latter shows a shallow minimum at $c/a = 1$. (The calculated bcc volume corresponds to a $\sim$3% reduction in lattice constant; the calculated fcc volume lies between these two values. The minima at $c/a = 1$ and $\simeq 0.87$ become more pronounced at still smaller lattice volumes.) Energies are relative to the low energy structure at each volume. Reproduced with the permission of the American Physical Society from Ref. [25].

which results in the expression

$$\delta U = 2\delta a \left\{ \frac{\partial U}{\partial a} - \frac{c}{a} \frac{\partial U}{\partial c} \right\}. \tag{3.5}$$

Now the extrema will occur when $\delta U = 0$, which in turn will depend on some particular relationships between $\partial U/\partial a$ and $\partial U/\partial c$. For elemental systems at bcc ($c/a = 1$) and fcc ($c/a = \sqrt{2}$) points, where the local symmetry (O_h) is higher than that at tetragonally distorted points, it is easy to that $\delta U = 0$, leading to an extremum. Explicit calculations carried out in this study (Ref. [25]) find that the nonequilibrium, fcc or bcc, structures in the middle of the $3d$, $4d$ and $5d$ transition metal rows are found to be local *maxima* in energy along the path of distortion, implying that they are *unstable* with respect to volume conserving distortions discussed above.

The above section underscores the usefulness of total energy calculations in dealing with lattice distortions of free standing metallic layers. For epitaxial multilayer systems, there are factors other than strain that can play important roles in stabilizing some of the unstable phases discussed above. Finite temperature effects, such as those due to vibrational and electronic entropy contributions, can sometimes turn out to be essential when determining stability between two closely lying metastable states. Vibrational entropy contributions to the free energy are generally considered more important than the electronic ones, although the latter contributions can sometimes be of the same magnitude as the entropy changes inferred from experiments [15,16]. In addition, the interface effects and associated energetics, discussed in the next section, can also be quite significant.

3.6.   Interfaces in metallic multilayers – Pb-Nb and Ag-Nb

With the advent of sophisticated experimental techniques, such as MBE that enable the growth of well defined metallic multilayer systems, it is becoming more and more important to understand the equilibrium thermodynamical properties related to the formation of interfaces. Electronic states at the interface are quite likely to be different from those found in either of the bulk metallic elements that form the interface and are likely to raise interesting experimental questions. For example, physical quantities such as interface energies and related bonding mechanisms are notoriously difficult to determine experimentally. However, with the present day total energy calculations (based on DFT), it has become possible to disentangle some of these and understand the underlying physics that is central to interface formation.

   As an illustrative example, consider the systems Pb-Nb and Ag-Nb [17]. The bulk phase diagrams of these two systems are vividly different. In Pd-Nb, the bulk phase diagram displays a tendency for strong compound formation, while in Ag-Nb, there is complete immiscibility, with no intermediate compounds. Figure 3.8 shows a tracing of the Pd-Nb binary alloy phase diagram, as reported in Ref. [18], as well as some calculated heats of formation for several assumed structures, PdNb, Pd_2Nb and Pd_3Nb. The heat of formation for the 50–50% compound, PdNb, is found below the line connecting the heats of pure Nb and Pd_3Nb, indicating that it is unstable and likely to phase separate into a mixture of Nb and Pd_3Nb at low temperature. In accordance with the phase diagram, Pd_2Nb in $MoPt_2$ structure and Pd_3Nb in $TiAl_3$ structure appear to be the most stable with heats of formation close to 0.5 eV, suggesting a preference for strong compound formation. For the Ag-Nb system, the phase diagram shows no intermediate compounds.

   If one follows the bulk phase diagrams naively, both these systems are not expected to favor pseudomorphic growth (i.e., growth of Pd on Nb or Ag on Nb does not appear likely). However, experimentally, it is found that Pd forms thick epitaxial layers on Nb(100) and at least a monolayer of Ag grows on Nb(110) [13]. This is contrary to the expected behavior, since for strong compound forming elements, one would expect interdiffusion, while for immiscible ones, clustering is to be expected.

   To understand the formation of surfaces and interfaces in layered structures, consider a substrate of element B, on which epitaxial layers of element A are grown. To discuss the energetics associated with such growth, we use ideas due to Gibbs as follows. The heat of adsorption in this case may be written as (Ref. [17])

$$\Delta E = \gamma_A - \gamma_B + \xi + \Delta E_{str}, \tag{3.6}$$

where γ_A and γ_B are the surface energies of the elemental surfaces, ξ is the bonding energy at the interface, and ΔE_{str} is the strain energy necessary to get A to be in registry with the substrate B. The strain and the surface energies are always positive (creation of surfaces and strains cost energy compared to the existing bulk solids), while the (interface) bonding energy, ξ, could either be positive (non-bonding) or negative (bonding), depending on the nature of the bonding orbitals at the interface. For a multilayer system

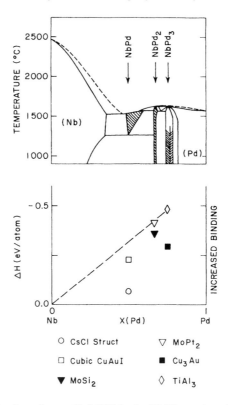

Figure 3.8.   A tracing of the phase diagram (Ref. [18]) for the $Pd_x Nb_{1-x}$ phase diagram plus calculated heats of formation for PdNb, Pd_2Nb, and Pd_3Nb for some ordered structures. Note that $MoPt_2$ and $TiAl_3$ are the observed structures at 2:1 and 3:1 compositions. Reproduced with the permission of the American Physical Society from Ref. [17].

ABAB ..., the above equation (Equation (3.6)) reduces to

$$\Delta E = 2\xi + \Delta E_{str}, \tag{3.7}$$

since there are no free surfaces. The factor of 2 accounts for the two interfaces present in a unit cell of this multilayer system. In order to avoid ambiguity in defining various contributions to Equations (3.6) and (3.7), we will define these as differences compared to bulk values. When $\Delta E \leqslant 0$, layer by layer growth is favored over clustering.

The strain energies were determined by calculating the energy cost of a given distortion in the elemental bulk system compared to bulk equilibrium crystal structure. Bulk superlattices (i.e., multilayers) ranging from Pd_2Nb_2 to Pd_6Nb_6, with the in-plane lattice constant constrained to that of Nb(001), all yielded negative heats of formation, as shown in Figure 3.9. However, for Pd-Nb(110) multilayers, the heats of formation become positive for Pd_3Nb_3 which implies that the (001) stacking is more energetically favorable when compared with the (110) stacking. There are several reasons for this: (1) the strain energy cost when trying to fit Pd into the Nb(001) face is half of that from the corresponding value for Nb(110); (2) the (001) face has more unlike neighbors (hence more bonding) compared to (110). The heats for alloy formation (carried out for various ordered alloys

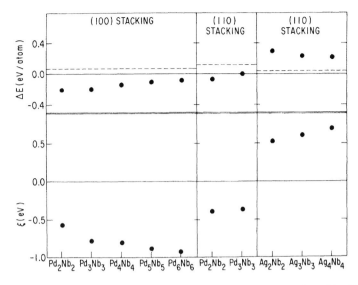

Figure 3.9. Top panel: the binding energies per atom, of $Pd_x Nb_x(001)$, $Pd_x Nb_x(110)$, and $Ag_x Nb_x(110)$ multilayers as a function of multilayer thickness. The dashed lines are the asymptotic ΔE for the infinitely thick multilayers. Bottom panel: ξ, the energy per atom pair in an interface obtained by subtracting the calculated distortion energies ΔE_{str}, for all Pd (or Ag) layers in a multilayer from the multilayer binding energy. Reproduced with the permission of the American Physical Society from Ref. [17].

of Nb-Pd in different stoichiometries and crystal structures) and bonding energies at the interface, ξ, are comparable, indicating that the multilayering can compete with the alloy formation. In Figure 3.10, the projected densities of states of the multilayer $Pd_6 Nb_6(001)$ show how the interface atoms differ from the others, in addition to other features [17].

However, for Ag-Nb, both alloy and multilayer heats are positive (see Figure 3.9), which is consistent with the bulk phase diagram which shows no intermediate compounds. The positive heats do not necessarily imply that Ag and Nb do not form bonds; breaking a Nb-Nb bond and forming a Ag-Nb bond results in no gain in energy. The multilayers, if formed, are likely to be metastable with kinetic processes not being able to take the system towards its global equilibrium configuration.

Some of the conclusions of this study are as follows: (1) For multilayer (ABAB ...) formation, the interface bonding energy should be comparable to the heats of formation of binary alloys of A and B; otherwise interdiffusion is likely to occur. (2) The surface energies must be comparable, so that there is no bias when A on B vs B on A deposition is considered. (3) ΔE_{str} should be small, i.e., there should be a reasonable lattice matching; otherwise, the positive and large strain energy will force the multilayer to collapse. Ideas similar to these could be developed for multilayers with more than two component elements. However, note that the above discussion is intended only as a guide to understanding multilayering using selected, nonmagnetic elements. Depending on their bulk phase diagrams, different binary systems may show different trends in multilayering. In magnetic systems, one has to take into account the energetics associated with magnetic ordering, resulting in more complex but rich physics.

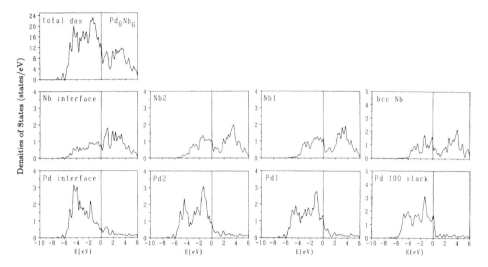

Figure 3.10.   The total density of states and local densities of states for $Pd_6Nb_6(001)$, where Nb2 refer to the Nb's adjacent to the interface Nb's and Nb1 to those once further removed. The right-hand panels display the density of states of bcc Nb and of Pd constrained to have Nb-like (001) layers and the Pd-Pd separation employed in all the $Pd_xNb_x(001)$ multilayer calculations. The total density of states involves a sampling over all crystal space while the lower panels have their sampling restricted to the charge within the atomic spheres.
Reproduced with the permission of the American Physical Society from Ref. [17].

3.7.   Magnetic 3*d* metals

Naturally occurring pure metals that are ferromagnetic, such as bcc Fe, fcc Ni and hcp Co, have played important and vital roles in the fundamental as well as applied aspects of magnetism, to say the least. For example, bcc Fe is the prototypical ferromagnet, known to man for ages. Other phases of Fe are known to exist at high pressure and temperature and these can be synthesized at room temperature (for example) under certain growth conditions. Metastability and advances in materials synthesis have paved the way for an unprecedented opportunity to examine and understand some long standing problems in (itinerant) magnetism. For example, fcc Fe has been predicted to be nonmagnetic, ferromagnetic, antiferromagnetic and spin-wave-like depending on its lattice constant (or growth conditions). It is also known to exist in a Cu matrix as nanoparticles (~50 nm) at low temperature. Again due to improvements in growth techniques among others, researchers have been able to grow different crystal structures of Fe, Ni and Co on suitable substrates under normal conditions, giving rise to different electronic and magnetic as well as other properties. Obviously the wide variety of magnetic structures and their competition under changes of physical variables lead to a better understanding of various forms of (itinerant) magnetism.

In iron, we encounter a rich phase diagram, consisting of the (ground state) bcc (α), a high temperature fcc(γ), and a high pressure hcp (ϵ) phase. For example, several years ago, the fcc-bcc transition in Fe was studied at some great length. There were several phases of fcc Fe such as paramagnetic, ferromagnetic and antiferromagnetic that were found to exist, depending on the volume per atom in the solid state. (See Figure 3.11 for

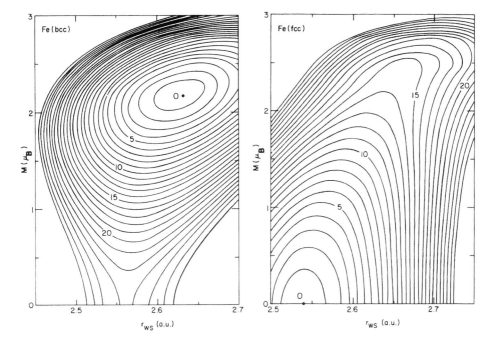

Figure 3.11. Energy contours using fixed moment method, giving rise to stable magnetic (bcc Fe) or nonmagnetic (fcc Fe) minima Reproduced with the permission of the American Physical Society from Ref. [19].

a theoretical search of stable phases of Fe). An expanded lattice favors the fcc phase. The interest in this system was renewed in the nineties after the neutron scattering experiments by Tsunoda [26] that were able to detect a spiral and noncollinear SDW (spin density wave). Later Hirai and others [27,28] looked at the wavevector dependence of the this state that was called a helical SDW.

3.8. Epitaxially grown magnetic systems

In this and the following sections, we will discuss various epitaxially grown, possibly magnetic, bilayer systems, where there is a reasonably close match of the lattice constants or corresponding (crystalline) faces of the two elements in question. This has been achieved experimentally using a substrate with an appropriate lattice match. Theoretically, there are predictions of magnetism in thin overlayers of some $4d$ and $5d$ transition metals (which are nonmagnetic in their normal bulk phases) on a suitable substrate (see Figure 3.12) [29]. In $3d$ metals where the outer d orbitals are compact, magnetism is more common when compared with $4d$ and $5d$ metals and epitaxial films provide a way to manipulate magnetic properties. One well studied example is the growth of thin fcc Fe films on Cu(100).

The fcc phase of Fe is not a naturally stable phase at room temperature, but can undergo pseudomorphic, epitaxial growth on suitably matched substrates. With a Cu substrate, the lattice mismatch is about 1%, where the mismatch is defined as the percentage dif-

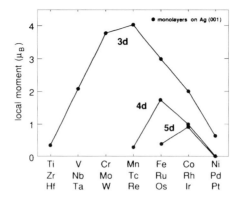

Figure 3.12.   Local moments of $3d$, $4d$, and $5d$ transition metal monolayers on Ag(001). Reproduced with the permission of the American Physical Society from Ref. [29].

ference in the relevant lattice parameters for the corresponding phase and the substrate (Equation (3.2)), on which epitaxial growth is expected. There are at least two important reasons for studying such growth. (1) New insights into atomic reorientation mechanisms and driving forces such as those during the first steps of martensitic nucleation. (2) Novel magnetic properties of such thin film systems which could be useful in various applications such as read/write devices.

As found in fcc Fe grown on Cu(100), if the substrate and the deposited material have the same crystal structure and there is a *reasonable* match in lattice parameters, (other substrates such as Ni or Cu_3Au share this property with fcc Fe), it is possible to grow a limited number of monolayers. These may be strained to some extent and this strain will be an important factor in determining the eventual (critical) thickness of the deposited layers; i.e., there will be a limiting thickness beyond which the epitaxial growth can not be sustained as discussed in Section 3.3. The deposited layers will try to relax back to a more energetically stable state through misfit dislocations or other, similar relaxation mechanisms.

It is worth mentioning a recent study, which compares the pseudomorphic growth of thin films of Fe on Cu(100) and Ni(100) substrates (Ref. [30]). Although the lattice mismatch in both cases is found to be about 1%, the strain is compressive with Ni and tensile in the case of Cu. The fcc-bcc transition is said to occur through different mechanisms in these two cases. With the Cu substrate, misfit dislocation-like ridge structures are seen to appear around 4.6 ML of Fe, which start transforming into bcc precipitates. With the Ni substrate, no misfit dislocation-like structures are observed and the transformation into bcc structure starts around 5.5 ML of Fe, by forming bcc island chains along the (001) direction with a significant misfit accommodating effect.

3.9.   More on epitaxially grown fcc Fe/Cu

The magnetic nature of fcc iron (γ-Fe) grown on Cu substrates has been of great interest, from both fundamental and applied perspectives for quite some time. This system exhibits a variety of magnetic phases and hence has become a unique case study. It does provide some useful insight into itinerant magnetic multilayers. Early experiments on this system showed ferromagnetic sheets coupled antiferromagnetically to each other with size-dependent Néel temperatures. Depending on the substrate crystal surface, i.e., whether it is (100) or (111), temperature or pressure, a structure with a different (magnetic) phase could be grown up to a few layer thickness.

For Fe films grown epitaxially on Cu(001) at low temperature, there seems to be agreement on the following points: (a) For thick films, the structure is bcc Fe, (b) below about 4 monolayers (ML), the structure is face centered tetragonal, (c) between 5 and 11 ML, an antiferromagnetic structure with ferromagnetic coupling between the surface and subsurface layers is found. There is disagreement about the nature of the magnetic coupling in the interior layers. It is instructive to go through the early first principles predictions and some experiments in chronological order, since a number of useful lessons can be learned by studying them.

Early theoretical calculations of thin fcc Fe slabs, having the lattice constant of Cu, were consistent in predicting that the surface (S) and the subsurface (S-1) layers are coupled ferromagnetically (Refs. [31–33]). While Fu *et al.* [31] and Fernando *et al.* [32] focused on thin (5–7 layers of) fcc Fe films, Kraft *et al.* [33] extended that work to a 11 layer film of fcc Fe. In all these studies, it was established that coupling beyond S-1 layers is antiferromagnetic (i.e., S-1 layer is coupled antiferromagnetically to S-2). Figure 3.13 is quite instructive; first, the nonmagnetic, layer projected DOS plots clearly show a high Fermi level state density at the surface, subsurface and center layers. Although naively, one would expect ferromagnetism in such a situation, here we encounter antiferromagnetism in addition to ferromagnetism; i.e., while each layer is ferromagnetic, coupling between the subsurface and center layers is antiferromagnetic. These properties can be tied to the properties of the Fermi surface, through a modified Stoner type mecha-

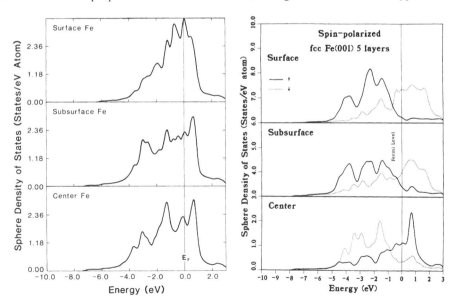

Figure 3.13.   Nonmagnetic and magnetic layer projected density of states of a 5 layer fcc Fe(001) film. Reproduced with the permission of the American Physical Society from Ref. [32].

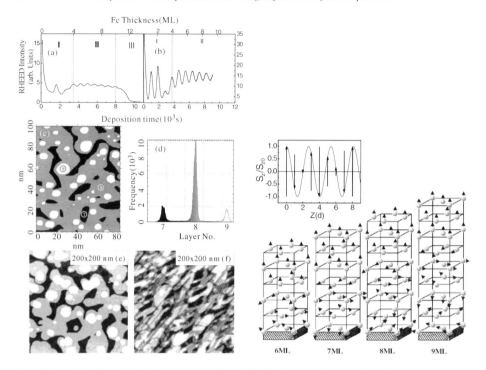

Figure 3.14.   Experimental work on fcc Fe(001) films indicating the complex magnetism in fcc Fe. Reproduced with the permission of the American Physical Society from Ref. [34].

nism. Also note that when the magnetic (layer) DOS are compared with the nonmagnetic (layer) DOS, there is a clear, *nonrigid* shift in the respective layer projected DOS (see Figure 3.13). One must note that none of these studies explored a possible, helical spin density wave formation. To search for such SDWs, a reasonably large unit cell has to be utilized and a (2×2) spin density matrix has to be incorporated, as discussed in Chapter 2 of this volume. The point related to S and S-1 ferromagnetism is very likely tied to the narrowing of the DOS in these layers and hence fulfilling a Stoner criterion. The decrease of the number of nearest neighbors near the surface, leads to bands that are more narrow compared to the bulk and the Cu lattice constant (expanded volume) provides more help in attaining a ferromagnetic surface.

Recent studies of this system have yielded new insights into the magnetic order in the interior of these fcc Fe films [34] (see Figure 3.14). Using MOKE, RHEED, LEED, STM and other techniques to characterize and monitor the growth of Fe layers on Cu(100), several key features of the structures were obtained. Polar and longitudinal MOKE measurements indicated the direction of the magnetic moments and how they orient themselves when cooling or warming the sample. It was concluded that a collinear type-I antiferromagnetic structure does not exist in these multilayer films, but a spin-wave like antiferromagnetism (AFM) is a possibility. Contrary to some predictions of a type-I AFM, the total moment for a 7 ML film of Fe on Cu(001) increased monotonically as the temperature decreased (for a 6 ML film, there is an additional step-like increase). An effective ordering temperature of about 200 K was found, below which the magnetic order persisted. Also, strong, reversible coercivity changes (increases) were observed with decreasing temperature. This was regarded as a fingerprint of turning on the exchange coupling between the top ferromagnetic layers and the interior antiferromagnetic layers. Dependence of the Curie temperature on the Cu cover layer in Cu/Fe/Cu(001) sandwiches has also been examined by Vollmer *et al.* [35].

More recent theoretical first principles studies, some of which have allowed lattice relaxations, have also demonstrated [36,37] that in ultra thin fcc Fe films, the surface and subsurface layers are ferromagnetically coupled. When the layer thickness is larger that 4, a non-collinear magnetic order has been found in the interior layers in Ref. [38], which implies that these films cannot be characterized by simple, uniaxial magnetic anisotropy.

In summary, thin fcc Fe films grown on substrates such as Cu, Ni or Cu$_3$Au have shown fascinating structural, magnetic and thermodynamic properties. This (metastable, pseudomorphic) system has provided clues with regard to the interplay among a variety of magnetic and nonmagnetic phases and as to how the magnetic properties are affected by the substrate. It has also provided structural information about the transition to the bcc phase of Fe (i.e., the ground state) when the number of overlayers exceeds a critical value.

3.10.   Epitaxially grown Fe$_{16}$N$_2$ films

For demanding, robust applications, such as magnetic recording discussed in a previous chapter that rely on magnetic properties, $3d$ transition elements such as Mn, Co and Fe can be described as the most straightforward candidates. However, there is an inevitable problem with these metals, namely they oxidize readily which can have a negative effect on their intended use. For the above reason, various compounds of these elements are of significant interest. Iron nitrides form such a class of compounds. In this section we will examine an epitaxially grown iron nitride, Fe$_{16}$N$_2$, where some of the previously discussed theoretical as well as experimental techniques have been utilized to unravel its complexities. The discussion below is based on some of the work reported in Ref. [39]. This compound, which has three inequivalent Fe sites, exemplifies the effects of lattice distortions, coordination numbers and similar changes in the local environment on the magnetic moments.

Iron nitride films provide an interesting as well as a controversial magnetic system. In 1972, Fe$_{16}$N$_2$ thin films obtained by evaporating iron in nitrogen were found to have unusually large saturated magnetizations [40]. Although these films did contain substantial amounts of α-Fe, which is the ground state bcc structure, the high magnetizations were attributed to the presence of the nitride phase. More recent experiments related to measuring the magnetic moments of Fe in such films have raised numerous questions, both experimental and theoretical.

Iron nitrides are fascinating examples of magnetic solids with a mixture of bcc- and fcc-like local iron environments subject to some distortions. We will consider two such nitrides, Fe$_4$N and Fe$_{16}$N$_2$, in this discussion. The ground state crystal structure of Fe$_4$N is a fcc lattice of Fe atoms with a nitrogen is a fcc lattice of Fe atoms with a nitrogen atom occupying the body center position; i.e., each nitrogen atom is surrounded by an octahedron of Fe atoms. There are two inequivalent Fe sites here, one (Fe-II) being closer to nitrogen than the other (Fe-I). Fe-I and N sites have O_h local symmetry while Fe-II

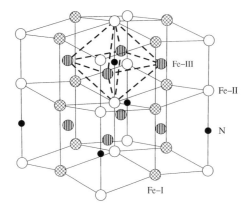

Figure 3.15.   Crystal structure of $Fe_{16}N_2$.

has D_{4h}. The (observed) lattice constant for this fcc structure is 3.80 Å. Figure 3.15 shows the crystal structure of $Fe_{16}N_2$. In this structure, there are three inequivalent Fe sites, two of them closer to being in a bcc environment. In Figure 3.15, Fe-I is labelled as the site furthest from N, while Fe-II is the closest to N. $Fe_{16}N_2$ can be thought of as alternate units of fcc Fe and Fe_4N units, with the atoms being allowed to relax (or having a unit cell consisting of 8 distorted bcc units). The primitive cell here is a body centered tetragonal one, and the local symmetries of Fe-I, Fe-II, Fe-III and N are D_{2d}, C_{4v}, C_{1h} and D_{4h} respectively. The lattice constants used for the body centered tetragonal $Fe_{16}N_2$ cell are $a = 5.72$ Å and $c = 6.29$ Å.

3.10.1.   Experimental background: Fe nitrides

There are several excellent reviews of the experiments [43,45]. During the 1990s, a number of different experimental groups have been able to grow thin films that contain $Fe_{16}N_2$ on GaAs substrates, using techniques such as Molecular Beam Epitaxy (MBE) [42]. The large polarization values (2.9 T) reported by the Hitachi group [41] for thin films containing $Fe_{16}N_2$ were the most notable due to their implications. It appears to be extremely difficult to obtain reasonably large samples of pure, single crystal $Fe_{16}N_2$. This is a major reason for the controversial claims with regard to average magnetization and, hence, Fe moments. Magnetization measurements have been done on samples that contain several phases such as bcc Fe, fcc Fe and Fe_4N, in addition to $Fe_{16}N_2$. Saturation magnetization σ_0 for such samples is always below the saturation value for pure, bcc iron. However, based on a phase analysis of the samples using either Mössbauer spectra or X-ray diffraction, it is possible to assign a σ_0 value for the $Fe_{16}N_2$ phase. This value ranges from 225–310 emu/g [43]. The upper end of these values lead to predictions of higher average moments and hence to an unusually large value of moment ($\geqslant 3.5\mu_B$) for the Fe site that is furthest from nitrogen. A more recent X-ray diffraction and Mössbauer study identifies the presence of disordered, octahedral N sites in a α'-$Fe_{16}N_2$ phase that co-exists with the ordered α'' phase, and their saturation magnetization values for the ordered phase have an upper bound of about 240 emu/g at room temperature.

The NMR work of Ref. [44] does not use this type of a phase analysis to deduce moments. It identifies the NMR frequencies (hence hyperfine fields) assigned to various sites and then relies on 'proportionality constants' that relate hyperfine fields to the values of magnetic moments at those sites. Using reasonable values for these 'constants', the NMR study appears to confirm the existence of a site moment that is $\simeq 3.5\mu_B$. One important point of agreement among these experiments is the observation of a large hyperfine field (of about 41.8 T for Fe-I site, compared to bcc α-Fe value of 33.8 T) for the $Fe_{16}N_2$ phase. Central to converting these to site moments are issues associated with site symmetry. Note that the experimental values for Fe_4N in Table 1 of Ref. [39] come from neutron diffraction studies [46] and comparison with Fe_4N's hyperfine fields lends credence to the value $3.5\mu_B$ for Fe-I in $Fe_{16}N_2$.

3.10.2. *Discussion of large Fe moments*

The usual Stoner parameters found for metallic bcc Fe are about 1 eV [47] (noting that I as defined in Ref. [47], is $U/2$). Figure 3.16 shows the values of the moments (found by self-consistently solving the Stoner–Hubbard Hamiltonian for a given U) as a function of U for bcc Fe, Fe_4N, and $Fe_{16}N_2$. The most striking result emerging from Figure 3.16 is that for U values near 1 eV, the Stoner theory derived moments and band structure

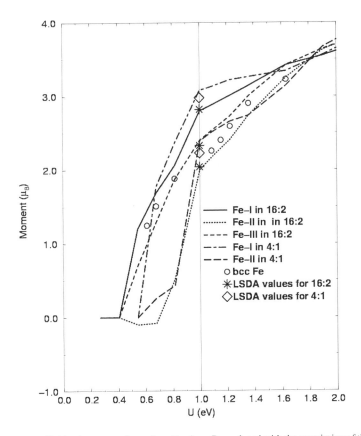

Figure 3.16.   Stoner–Hubbard moments for various U values. Reproduced with the permission of the American Physical Society from Ref. [39].

results show good agreement. This is true for bcc Fe and also for the two nitrides under study. More importantly, this version of Hubbard–Stoner theory indicates that the U values have to be unusually large, namely $\simeq 2$ eV, in order to have site moments that are around $3.5\mu_B$, for Fe_4N and $Fe_{16}N_2$. Interestingly, even bcc Fe will produce similarly large magnetic moments at such high U values. All the Fe sites in the two nitrides would carry comparable moments at such large U values due to the high degree of saturation.

In order to obtain a saturation moment of $3.5\mu_B$ for the iron sites under consideration, one has to resort to U values that are measurably larger than the ones attributed to ordinary metallic iron; this may, in turn, imply somewhat unphysical charge transfers between sites. It would appear that, even with the Hubbard–Stoner approximation employed here, one is hard pressed to rationalize magnetic moments of the size which have been argued experimentally unless strong correlation effects are accommodated in these metallic systems.

3.11.   More on Fe/Cr

In the multilayer system Cu/Co, the metal that is magnetic in its (bulk) ground state is Co, while Cu is nonmagnetic in its ground state fcc structure. When both components in such a multilayer system have (bulk) ground state structures that are magnetic, it is natural to expect more complex magnetic properties (such as a spin density wave (SDW)) in the resulting structure. This is exactly what is observed in the multilayer system Fe/Cr, which was previously discussed briefly in Chapter 2. Such magnetic structures are likely to depend on the crystallographic orientations of the two systems as well as the interface driven effects. There are excellent review articles on this system (see, for example, Refs. [48–50]). Below, we summarize some of the salient features of the multilayer system, Fe/Cr, discovered mostly during the 1990s.

Early experimental work (prior to 1995) on this system consisted of using techniques Scanning Electron Microscopy with Polarization Analysis (SEMPA), Brillouin Light Scattering (BLS), Ferromagnetic Resonance (FMR) and MOKE, which were unable to directly determine the existence of a SDW. However, notable SEMPA experiments in the early 1990s, carried out at NIST [51], measured the exchange coupling of a Cr wedge grown on a Fe whisker with a thin Fe layer on top of Cr. This coupling was seen to oscillate (flip) between F and AF, with a periodic phase slip. In 1995, Polarized Neutron Reflectometry (PNR) in combination with SMOKE was used, for the first time, to confirm the existence of a non-collinear magnetic behavior, i.e., the SDW [52]. The lattice mismatch in (001) Fe and Cr is small, allowing near perfect interfaces. First principles and tight binding calculations, which assume the existence of perfect interfaces, predict that the SDW loses its rigidity within a couple of layers.

Fe/Cr system has been the subject of numerous controversies since the discovery of the GMR effect [50]. The spacer layer is Cr and a SDW was observed in thin Cr films sandwiched by Fe layers. There appears to be a rich variety of possibilities for the Fe/Cr multilayers, as far as the SDW is concerned. Depending on the alignment of the adjacent Fe layer moments, it is claimed that a collinear or non-collinear (helical) SDWs can be stabilized. When the adjacent Fe layer moments have either FM or AFM coupling, a collinear SDW is found (see Ref. [53]) in first principles calculations while if they are at, say 90 degrees, a helical SDW is deemed possible (see Ref. [50]). Understanding the effect of the SDW on the interlayer coupling has been difficult, to say the least.

3.12. bcc Nickel grown on Fe and GaAs

Nickel is another $3d$ transition metal that has been extensively studied in its face centered cubic phase, which is the naturally occurring, stable phase. Similar to iron and cobalt, there have been numerous attempts to grow different phases of nickel, using pseudomorphic growth on an appropriate substrate. Such studies of metastable phases pave the way for a better understanding of itinerant magnetism, enhanced moments etc. However, it has been difficult to grow (non-naturally occurring) bcc nickel. Moruzzi *et al.* had theoretically predicted that bcc Ni at equilibrium would be paramagnetic (with a lattice constant of 2.773 Å), but a ferromagnetic transition would occur with an expanded lattice constant of 2.815 Å (at an expansion of about 1.5%).

At the bcc Fe lattice constant, bcc Ni shows an expansion of about 3% compared to the calculated equilibrium value and hence is expected to be ferromagnetic. Early experimental studies of bcc nickel by Guo *et al.* [54] and Heinrich *et al.* [55] were done on an iron substrate, namely Fe(001). Both groups agreed that bcc nickel could be grown up to about 6 layers and thicker layers showed a more complicated, modified bcc structure. Such thin films were found to be ferromagnetic with moments of ranging from 0.4 to $0.8\mu_B$ per nickel atom with some conflicting reports of Curie temperature (Refs. [56–58]). In the spin polarized photoemission study of Brookes *et al.* [58], it was shown that bcc Ni grown on Fe(001) is ferromagnetic at room temperature with bands exchange split by about 0.3 eV.

The growth of nickel on a magnetic surface such as Fe(001) is bound to have nontrivial effects on its magnetic properties, at least close to the interface region. Since the (pure) bcc Ni system is ultra thin, one cannot rule out the possibility of iron affecting the magnetism of Ni (such as enhancing Ni moments). A more recent experiment reports the growth of bcc Ni on a GaAs substrate [59], whereby the effects of magnetism due to the substrate can be ruled out. Their results can be summarized as follows: Magnetically isolated bcc Ni, grown on GaAs(100) up to thicknesses less than 3.5 nm, shows a Ni moment of $0.52\pm 0.08\mu_B$ with a Curie temperature of 456 K (in contrast to bulk fcc Ni Curie temperature, which is 627 K). The cubic anisotropy was measured to be $K_1 = 4.0 \times 10^5$ ergs-cm^{-3}. The positive value of K_1 implies that bcc Nickel has the same easy axis of magnetization as bcc Fe, which is the opposite to that of fcc Ni. The energy band structure for bcc Ni (determined using ARPES) appears to be different along the (100) direction, and this is mentioned as a reason for the observed differences in the magnetic anisotropies between fcc and bcc Ni in the above study.

3.13. Summary

The ability to synthesize artificial structures has undergone revolutionary advances over the past few decades. Epitaxial growth of layered structures has played an important role in this revolution. Understanding and utilizing such artificial growth processes can pose many challenges, both experimental and theoretical; in this chapter some of the important concepts, related to pseudomorphic growth, phase stabilities, magnetic structures have been addressed briefly. These systems are quite fascinating, at least in part, due to numerous novel properties that are not found in their bulk counterparts.

References

[1] F.C. Frank, J.H. van der Merwe, Proc. Roy. Soc. A **198**, 205 (1949).

[2] M. Volmer, A. Weber, Z. Phys. Chem. **119**, 6274 (1981).

[3] J.N. Stranski, L. Krastanov, Ber. Akad. Wiss. Wien **146**, 797 (1938).

[4] M. Wuttig, X. Lu, Ultrathin Metal Films (Springer-Verlag, Berlin–Heidelberg, 2004).

[5] D.R. Frankl, J.A. Venables, Adv. Phys. **19**, 409 (1970).

[6] M.V.R. Murty, B.H. Cooper, Phys. Rev. Lett. **83**, 352 (1999).

[7] M. Kalff, G. Cosma, T. Michley, Phys. Rev. Lett. **81**, 1255 (1998).

[8] U. Grandmann, in: Handbook of Magnetic Materials, vol. 7, Elsevier, Amsterdam (1993).

[9] R. Vollmer, T. Gutjahr-Löser, J. Krischner, S. van Dijken, B. Poelsima, Phys. Rev. B **60**, 6277 (1999).

[10] N. Weber, K. Wagner, H. Elmers, J. Hauschild, U. Grandmann, Phys. Rev. B **55**, 14121 (1997).

[11] S. Murphy, D. Mac Mathuna, G. Mariotto, I.V. Shvets, Phys. Rev. B **66**, 195417 (2002).

[12] H. Elmers, J. Hauschild, H. Höche, U. Grandmann, H. Bethge, D. Heuer, U. Köhler, Phys. Rev. Lett. **73**, 898 (1994).

[13] M.J. Sagurton, M. Strongin, F. Jona, J. Colbert, Phys. Rev. B **28**, 4075 (1983); M.W. Ruckman, L.Q. Jiang, Phys. Rev. B **38**, 2959 (1988).

[14] M. Rasamny, M. Weinert, G.W. Fernando, R.E. Watson, Phys. Rev. B **64**, 144107 (2001).

[15] C. Zener, Phys. Rev. **71**, 846 (1947); J. Fridel, J. Phys. (Paris) Lett. **35**, L59 (1974).

[16] R.E. Watson, M. Weinert, Phys. Rev. B **30**, 1641 (1984).

[17] M. Weinert, R.E. Watson, J.W. Davenport, G.W. Fernando, Phys. Rev. B **39**, 12585 (1989).

[18] W.G. Moffart, Handbook of Binary Phase Diagrams (Genium, Schenectady, NY, 1978).

[19] V.L. Moruzzi, P.M. Marcus, K. Schwartz, P. Mohn, Phys. Rev. B **34**, 1784 (1986); V.L. Morruzi, P.M. Marcus, in: K.H.J. Buschow (Ed.), in: Handbook of Magnetic Materials, vol. 7, North-Holland, Amsterdam (1993).

[20] E.C. Bain, Trans. Am. Inst. Min., Metall. Pet. Eng. **70**, 25 (1924).

[21] M.J. Mehl, L.L. Boyer, Phys. Rev. B **43**, 9498 (1991).

[22] M.J. Mehl, A. Aguayo, L.L. Boyer, Phys. Rev. B **70**, 014105 (2004).

[23] F. Jona, P.M. Marcus, Phys. Rev. B **63**, 094113 (2001).

[24] F. Jona, P.M. Marcus, Phys. Rev. B **65**, 155403 (2002).

[25] P.J. Craievich, M. Weinert, J.M. Sanchez, R.E. Watson, Phys. Rev. Lett. **72**, 3076 (1994); P.J. Craievich, J.M. Sanchez, R.E. Watson, M. Weinert, Phys. Rev. B **55**, 787 (1997).

[26] Y. Tsunoda, N. Kunitomi, R.M. Nicklow, J. Phys. F **17**, 2447 (1987);
 Y. Tsunoda, J. Phys. C **1**, 10427 (1989);
 Y. Tsunoda, Y. Nishioka, R.M. Nicklow, J. Magn. Magn. Matter **128**, 133 (1993).

[27] K. Hirai, Prog. Theo. Phys. Suppl. **101**, 119 (1990).

[28] M. Uhl, L.M. Sandratskii, J. Kübler, J. Magn. Magn. Matter **103**, 314 (1992).

[29] S. Blügel, Phys. Rev. Lett. **68**, 851 (1993).

[30] J. Shen, C. Schmidthals, J. Woltersdorf, J. Kirschner, Surf. Sci. **407**, 90 (1998).

[31] C.L. Fu, A.J. Freeman, Phys. Rev. B **35**, 935 (1987).

[32] G.W. Fernando, B.R. Cooper, Phys. Rev. B **38**, 3016 (1988).

[33] T. Kraft, P.M. Marcus, M. Scheffler, Phys. Rev. B **49**, 11511 (1994).

[34] D. Qian, X.F. Jin, J. Barthel, M. Klaua, J. Kirschner, Phys. Rev. Lett. **87**, 227204 (2001).

[35] R. Vollmer, S. van Dijken, M. Schleberger, J. Kirschner, Phys. Rev. B **61**, 1303 (2000).

[36] Xilin Yin, Klaus Hermann, Phys. Rev. B **63**, 115417 (2001).

[37] R.E. Camley, Dongqi Li, Phys. Rev. Lett. **84**, 4709 (2000).

[38] R. Lorenz, J. Hafner, Phys. Rev. B **58**, 5197 (1988).

[39] G.W. Fernando, R.E. Watson, M. Weinert, A.N. Kocharian, A. Ratnaweera, K. Tennakone, Phys. Rev. B **61**, 375 (2000).

[40] T.K. Kim, M. Takahashi, Appl. Phys. Lett. **20**, 492 (1972).

[41] Y. Sugita, K. Mitsuoka, M. Momuro, H. Hoshiya, Y. Kozono, M. Hanazono, J. Appl. Phys. **90**, 5977 (1991).

[42] H. Takahashi, M. Mirusoka, M. Komuro, Y. Sugita, J. Appl. Phys. **73**, 6060 (1993);
 M. Kumuro, Y. Kozono, M. Hanazono, Y. Sugita, J. Appl. Phys. **67**, 5126 (1990).

[43] R.M. Metzger, X. Bao, M. Carbucicchio, J. Appl. Phys. **76**, 6626 (1994).

[44] Y.D. Zhang, J.I. Budnick, W.A. Hines, M.Q. Huang, W.E. Wallace, Phys. Rev. B **54**, 51 (1996).

[45] J.M.D. Coey, Phys. World **6**, 25 (1993);
 J.M.D. Coey, J. Appl. Phys. **76**, 6632 (1994).

[46] B.C. Frazer, Phys. Rev. **112**, 751 (1958).

[47] J.F. Janak, Phys. Rev. B **16**, 255 (1977).

[48] D.T. Pierce, J. Unguris, R.J. Celotta, M.D. Stiles, J. Magn. Magn. Mater. **200**, 290 (1999).

[49] H. Zabel, J. Phys. Cond. Matter **11**, 9303 (1999).

[50] R.S. Fishman, J. Phys. Cond. Matter **13**, R235–R269 (2001).

[51] J. Unguris, R.J. Cellota, D.T. Pierce, Phys. Rev. Lett. **67**, 140 (1991);
 J. Unguris, R.J. Cellota, D.T. Pierce, Phys. Rev. Lett. **69**, 1125 (1992).

[52] A. Schreyer, J.F. Ankner, Th. Ziedler, H. Zabel, M. Schäfer, J.A. Wolf, P. Grünberg, C.F. Majkrzak, Phys. Rev. B **52**, 16066 (1995).

[53] K. Hirai, Phys. Rev. B **59**, R6612 (1999).

[54] G.Y. Guo, H.H. Wang, Chin. J. Phys. **38**, 949 (2000).

[55] B. Heinrich, et al., J. Vac. Sci. Technol. A **4**, 1376 (1986);
 B. Heinrich, et al., J. Crystal Growth **81**, 562 (1987).

[56] J.A.C. Bland, et al., J. Magn. Magn. Mater. **93**, 331 (1991).

[57] Z. Celinski, et al., J. Magn. Magn. Mater. **166**, 6 (1997).

[58] N.B. Brookes, A. Clarke, P.D. Johnson, Phys. Rev. B **46**, 237 (1992).

[59] C.S. Tian, et al., Phys. Rev. Lett. **94**, 137210 (2005).

Chapter 4

Magnetic Anisotropy in Transition Metal Systems

4.1. Basics of magnetic anisotropy

From the early periods of civilization, magnetic anisotropy has been made use of in man-made devices, such as magnetic compasses. In such a compass, it is well known that the magnetic needle points along a specific direction, determined by an externally applied magnetic field. Magnetic anisotropy is due to the electronic exchange forces, which can be quite significant in magnetic materials. Even in the absence of external magnetic fields, these forces are strong enough to align magnetic moments present in a given material. Usually an external field is necessary to get these moments to align along, even in materials that are likely to be strongly ferromagnetic. The reason for such behavior is the presence of magnetic domains or subvolumes. Each one of these domains could possess a saturated (i.e., maximum possible) moment, but different domains may not be aligned with each other. This non-alignment is the cause of unsaturation in typically ferromagnetic materials. An applied (magnetic) field of sufficient strength is necessary to bring such domains to align themselves.

If one considers an array of spins $\mathbf{S}_i$, coupled by a Heisenberg type interaction

$$J\mathbf{S}_i \cdot \mathbf{S}_j,$$

they would exhibit rotational symmetry, i.e., a rigid rotation of the spins would leave the Hamiltonian invariant. However, an external magnetic field can be used to break this rotational symmetry so that the spins would have a preferred direction. Even in the absence of such an external field, real materials usually have a preferred direction. In a given ferromagnetic, transition metal, there is a so-called "easy axis of magnetization", which is identified as its preferred axis of magnetization. The Magnetic Anisotropy Energy (MAE) E_{ani} is the energy associated with the orientation of the magnetic moments in a condensed matter system. For cubic systems, the leading term in anisotropy breaks the lattice (local) symmetry (i.e., there should be no $l = 2$ term in the energy expression of a cubic system) and this anisotropy energy may (sometimes) be expressed as

$$E_{ani} = K_1 \sin^2 \theta + K_2 \sin^4 \theta, \tag{4.1}$$

where θ is the polar angle with respect to the surface normal. The coefficients K_1 and K_2 depend on the system as well as temperature. The MAE is determined through how the magnetic moments are coupled to one another as well as to the lattice. This could come from either the spin–orbit coupling or the magnetic dipolar interaction. At low temperature, alignment of magnetization along the easy axis corresponds to a lower MAE compared to alignment along other axes. However, by applying a relatively strong external magnetic field, it is possible to change this orientation (or alignment). Magnetic anisotropy can be linked with switching the orientation of magnetization of a given magnetic material from its easy axis towards its hard axis. The energy required to change this orientation is identified as the magnetic anisotropy energy. Usually, this energy is found to be "small", i.e., of the order of a few μeVs (or smaller) per atom in 3-dimensional systems. However, in lower dimensional systems, such as multilayers and nanoparticles, this energy could be higher by a few order of magnitudes. The dipolar term is usually called the shape anisotropy, and could be substantial when the surface to volume ratio is high. We discuss below both these terms in some detail.

4.1.1.   Magnetocrystalline anisotropy (MCA)

This is the magnetic anisotropy introduced by the coupling of the lattice (orbital degrees) with electronic spin, i.e., by the spin–orbit interaction. Magnetocrystalline anisotropy (MCA) energy is identified as the energy associated with the spin–orbit interaction in transition metal systems. The magnetic field generated by the orbital motion of the electrons interacts with the spins and the system looks to minimize this energy, by choosing an "easy axis of magnetization". In Figure 4.1, various possible axes of magnetization are shown as an illustrative example for a fcc crystal. Table 4.1 shows a classification of magnetic materials according to easy directions of magnetization with some standard examples. From an itinerant (or band structure) point of view, the spin–orbit split, one particle energy levels near the Fermi energy are responsible for the lowering of energy. When a state having a certain weight for an orbital component $+m$ is occupied while the corresponding one with orbital component $-m$ is unoccupied (or has a different weight of occupancy), there is a preferred direction, at least for the orbital moment. When coupled with the spin degrees, the total moment is likely to be pointed along a particular direction in space in order to minimize energy, i.e., the degeneracy of the levels with total angular momentum $j = l \pm 1/2$ gets broken here.

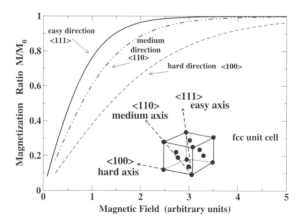

Figure 4.1.   Various possible axes of magnetization in a cubic fcc crystal. In this example (111) is the easy axis while (100) is the hard axis.

Table 4.1 Classification of magnetic materials according to easy directions of magnetization [1]

Type	Class	Easy direction	Examples
I	Uniaxial	One easy axis	Hexagonal, orthorhombic, tetragonal materials with positive anisotropy
II	Planar	Two or more easy axes in a plane	Hexagonal, tetragonal materials with negative anisotropy
III	Multiaxial	Three or more nonplanar easy axes	Cubic, metallic glasses, polycrystalline materials with small grains

Positive/negative anisotropy is simply a way of tagging anisotropy constants as in Figure 4.3.

Experimentally, these anisotropy measurements have yielded results that need careful consideration and analysis. For example, X-ray Magnetic Circular Dichroism (XMCD) measurements in Co/Pd and Co/Pt have shown significant L_z (orbital moment along the surface normal) values (≈ 0.29 and $0.21 \mu_B$) compared to clean hcp Co films ($\approx 0.13 \mu_B$). A simple relationship between L_z and the measured anisotropy was not found; for example, although the anisotropy constant K changes significantly with thickness t of the Co film, L_z was found to be insensitive to t [2]. (See also the section on XMCD in this chapter.)

4.1.2. Early work

In a classic paper published in 1937, van Vleck [3] addressed the issue of magnetic anisotropy using a pair interaction term. This was carried out well before our current understanding of itinerant magnetism. He pointed out that the orbital interaction energies between two atoms depend not only on the relative orientations of the orbital moments but also the direction of the line joining the two atoms. This is in contrast to a Heisenberg type spin–spin interaction which is isotropic with respect to rigid spin rotations, hence the spin–spin exchange energy has no preferred spatial axes, as those listed in Table 4.1, since there is no coupling to orbital degrees of freedom. In van Vleck's paper, the orbital dependence is referred to as being due to "orbital valence" and a resulting interaction between the atoms. This problem is somewhat subtle and is further discussed in Chapter 2 of this volume where orbital effects in atoms are considered. Later, Néel [4] extended van Vleck's model to surfaces showing that the reduced symmetry at a surface gives rise to anisotropies that are significantly different from bulk values. For example, he was able to derive a simple expression for the anisotropy energy,

$$E_{ani} = -K_S \cos^2 \theta, \tag{4.2}$$

with the anisotropy constant K_S differing for (111) and (100) surfaces (θ being the polar angle with respect to the surface normal). In Néel's pair anisotropy model [4], the energy is a sum of nearest neighbor pair interactions that depend on the local magnetizations as well as the vector that connects the two sites, which can be expressed as a sum of

Legendre polynomials P_l.

$$E_{\text{Neel}} = \sum_{i,j} \left\{ A(r_{ij}) P_2(\phi_{ij}) + B(r_{ij}) P_4(\phi_{ij}) + C(r_{ij}) P_6(\phi_{ij}) + \cdots \right\}. \tag{4.3}$$

This expression can be simplified using the symmetries of the local environment. For example in a cubic environment, the sum involving the P_2 term vanishes. However, if and when it survives (when the symmetry is low enough), this uniaxial term could be nonzero. For example, in the presence of stepped surfaces in fcc and bcc crystals, this term does survive. The next two higher order terms can be expressed in a (cubic) fcc lattice as

$$K_1 \left(\alpha_x^2 \alpha_y^2 + \alpha_y^2 \alpha_z^2 + \alpha_z^2 \alpha_x^2 \right) + K_2 \left(\alpha_x^2 \alpha_y^2 \alpha_z^2 \right), \tag{4.4}$$

where the first (K_1) term represents the sum of P_4 terms and the second (K_2) term is a linear combination of P_4 and P_6.

Although the pair interaction model yields useful insights, it is not sufficient to provide quantitatively reliable estimates for magnetic anisotropy. The reason is quite clear. If symmetry broken states and their relative occupations are involved in the determination of anisotropy, then, without explicit occupied state (and Fermi level) information, it would be very difficult to accurately estimate such a difference in energies. Simple pair interaction terms do not carry such information.

Examining Hartree–Fock type exchange integrals and nonspherical terms therein shows how orbital polarizations arise. More importantly, evaluation of the spin–orbit energy term along various directions of spin quantization in a lattice yields useful information about magnetic anisotropy. As discussed above, in high symmetry cubic structures, this correction occurs in the fourth order (the lowest lattice harmonic corresponds to $l = 4$ here). When this symmetry is lowered, such as at a surface, lower order terms become nonzero and the anisotropy terms get larger. Since the ferromagnetism in most $3d$ elements arises from spins (or spin-dominant), it may not be obvious as to why the orbital polarization should play any significant role. However, the spin–orbit coupling term will necessarily bring a (spatial) directional dependence to the energy discussed above.

The coupling of AF thin films with ferromagnetic layers has become an important topic due to their relevance in the so-called "exchange-bias" effect. In this context, a rich variety of AF-FM structures and properties have been studied since it was discovered in the 1950s [5–7]. Numerous magnetic devices make use of the exchange-bias effect which brings about a unidirectional magnetic anisotropy at the interface of a FM/AFM sandwich. Such a unidirectional anisotropy can be obtained by growing an antiferromagnetic layer on top of a magnetically saturated ferromagnetic layer or by field-cooling the system from above its blocking temperature T_B (defined as the temperature at which the exchange-bias or the shift vanishes). This topic is further discussed later in this chapter.

4.1.3.   Dipole–dipole interaction and related anisotropy

Another important contribution to magnetic anisotropy arises from the long-range dipole–dipole interaction. For example, the dipolar term is believed to be the dominant anisotropy term in MnF_2, which has a rutile structure and tetragonal symmetry with (001) uniaxial anisotropy [8]. The total magnetic (dipole) moment in an atom clearly interacts with other

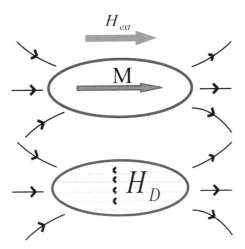

Figure 4.2. Demagnetization field H_D induced by the surface dipole distribution, which is due to finite size of the sample.

dipole moments and the resulting classical (Maxwellian) interaction leads to the dipole–dipole energy term. This term is nothing but a discrete sum of dipole–dipole interaction energies from dipole moments $\mathbf{m}_i$ and $\mathbf{m}_j$.

This discrete sum representing dipole–dipole interaction energy between two dipoles,

$$\sum_{\substack{i,j \\ i \neq j}} \frac{\mathbf{m}_i \cdot \mathbf{m}_j}{r_{ij}{}^3} - \frac{3\{\mathbf{m}_i \cdot (\mathbf{r}_j - \mathbf{r}_i)\}\{\mathbf{m}_j \cdot (\mathbf{r}_j - \mathbf{r}_i)\}}{r_{ij}{}^5}, \tag{4.5}$$

may be written (in the continuum limit) as an integral,

$$F_{shape} = -\mu_0/(2V) \int d^3r\, \mathbf{M}(\mathbf{r}) \cdot \mathbf{H}_D(\mathbf{r}). \tag{4.6}$$

The field, $\mathbf{H}_D(r)$, is the magnetic field due to all the other dipoles around the dipole with moment $\mathbf{M}(\mathbf{r})\, d^3r$ and is sometimes referred to as the demagnetizing field. This shape anisotropy term is divergent for an infinite sample and it is the finite size (and shape) of the sample that leads to a convergent value. Hence, the shape anisotropy energy is very much dependent on the boundaries and the direction determined by minimizing this dipolar energy will be the easy axis of magnetization, if this alone were to determine such an easy direction. For a few monolayers, the above continuum result cannot be valid and the discrete sum (in Equation (4.5)) has to be evaluated carefully. This discrete sum can be broken into surface and bulk contributions. Figure 4.2 is a simple illustration of the origin of the demagnetization field.

4.1.4. Magnetoelastic anisotropy

In a magnetic material that is under an external stress (or strain) is likely to undergo changes in magnetic anisotropy. This is a result of changes in the lattice parameters induced by the external stress and is appropriately referred to as magnetoelastic anisotropy.

The inverse of this process, i.e., the changes in lattice parameters due to magnetization, is usually labeled "magnetostriction". The associated elastic energy in an isotropic medium is

$$E_{mag\text{-}el} = \frac{3}{2}\lambda_m \sigma \cos^2 \theta, \tag{4.7}$$

where σ denotes the stress and λ_m is the magnetostriction constant with θ denoting the angle between magnetization and the direction of (uniform) stress. Magnetostriction λ can be either positive or negative and the magnetoelastic term can sometimes be an important contribution to the overall magnetic anisotropy.

There could be many reasons for a stress/strain to develop in magnetic multilayers. Epitaxial pseudomorphic growth usually leads to such strains. In addition, thermal expansion results in strains altering the magnetoelastic anisotropy.

4.1.5. Perpendicular magnetic anisotropy

In view of its implications for the next generation magnetic recording devices, perpendicular magnetic anisotropy (PMA) is an area of research that has attracted a significant amount of interest. The experimental studies of PMA in magnetic films and multilayers can be traced back to the work of Carcia *et al.* [9] on Pd/Co thin film, layered and coherent structures. In their work, Carcia *et al.* used a sandwich structure of alternating layers of Pd with thin Co layers. When the thickness of Co film was less that 8 Å, the magnetization of Co was found to be in plane in this multilayer system. They argued that this was due to interface and strain anisotropies. The effective anisotropy field H_K was obtained from the extrapolated intersection of the $H_{\parallel}$ magnetization curve with the saturation value of the $H_{\perp}$ curve, and found to depend linearly on $1/t$ where t is the thickness of the Co film. This linear dependence in single magnetic thin films is sometimes regarded as evidence for the presence of surface or interface anisotropy (see Equation (4.8)).

When analyzing bulk (volume), surface and interface contributions to anisotropy of a single (atom type) layer, it is customary to express the anisotropy constant in terms of volume, surface and interface related constants (phenomenologically) as

$$K_{eff} = K_v + \frac{K_s + K_{int}}{t}, \tag{4.8}$$

where t denotes the film thickness (i.e., tK_{eff} depends linearly on the thickness in this model with the slope given by K_v). Equation (4.8) is a simple expression which examines how the thickness affects various contributions to the effective anisotropy. For some layered systems the above expression provides a reasonable description of the variation of anisotropy with thickness. Figure 4.3 shows a possible scenario where, at large multilayer thicknesses ($t \to \infty$), the magnetization becomes parallel to the film, while at small t, it stays perpendicular to the plane. Here the volume contribution K_v is negative, favoring in-plane magnetization, while the y-intercept, representing surface/interface contributions, is positive, favoring perpendicular magnetization. There is a critical thickness $t_{\perp}$ at which the magnetization changes its direction. Although this form has been used successfully to interpret numerous experiments, there could be problems tied to the linear dependence and the separation into volume and surface/interface contributions as mentioned above;

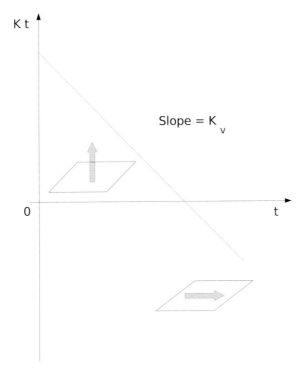

Figure 4.3. Possible variations of magnetization directions and Kt as a function of multilayer thickness t. There is a critical thickness at which the magnetization changes its direction. Although this form is often used to analyze volume and surface contributions, it has to be interpreted with care.

the effective anisotropy is likely to have a more complicated behavior, especially at small thicknesses.

In Ref. [9], as well as many other similar studies, the multilayers consist of a nonmagnetic metal (such as Pd) coated with magnetic layers of another metal (such as Co). If both layers are ferromagnetic, then Equation (4.8) has to be modified to include shape and volume anisotropies due to the second set of ferromagnetic layers. For example, in Co/Ni multilayers [10,11] the thicknesses of both Co and Ni become important and total anisotropy energy was obtained from the difference between the parallel and perpendicular magnetization energy fitted into the form

$$\lambda K_u = \frac{1}{1+\alpha_r}\left(K_v{}^{Co} + \alpha_r K_v{}^{Ni}\right)\lambda + 2K_s{}^{Co/Ni}. \tag{4.9}$$

Here $\alpha_r = t_{Ni}/t_{Co}$ is the thickness ratio. Interface and volume contributions can be evaluated by conducting two experiments, one at a fixed Co thickness by varying t_{Ni} and the other at a fixed Ni thickness by varying t_{Co}. For $\alpha_r \simeq 2$, an interface contribution $K_s = 0.31$ mJ/m^2 and an averaged volume contribution of $K_v = -0.39 \times 10^6$ J/m^3 were obtained (Refs. [10,11]) indicating perpendicular magnetization for $\lambda < 15$ Å and parallel magnetization above 15 Å (i.e., negative values of K_v favoring in plane magnetization).

The verified linear behavior of Equation (4.9) indicates that the interface contribution can be separated out at small as well as large thicknesses from the volume contribution.

In the Co/Pd study of Carcia *et al.* [9], using a linear fit to data, it was concluded that the critical Co thickness was $\simeq 8$ Å for the effective anisotropy to vanish. From the slope of a plot, similar to the above, the magnitude of K_v was found to be 0.9×10^6 J/m^3, while the magnitude of K_s was found from the intercept to be 0.16 mJ/m^2 favoring perpendicular magnetization; however, the magnitude of K_v was higher than the bulk crystallographic value for hcp Co, which was possibly due to magnetostrictive effects. A tensile strain in Co (arising from the lattice parameter mismatch with Pd) was thought to be responsible for this large value. Also, they were able to rule out significant enhancements of induced magnetization in Pd. A comparison of the Co/Pd and Co/Ni work mentioned here shows that the anisotropy constants and critical thicknesses very much depend on the multilayer system that is being investigated; it is also evident that bulk anisotropy parameters can get significantly modified at surfaces and interfaces.

4.1.6.   XMCD and XMLD for anisotropy measurements

As discussed in Chapter 5, X-ray magnetic circular dichroism (XMCD) has been established as a reliable technique to determine element specific, orbital and spin moments. There are sum rules that yield orbital and spin moments in a straightforward, quantitative fashion. The spins tend to align along an easy axis of the crystal in order to minimize the MCA energy. This is driven by the orbital degrees as demonstrated in experiments using Co films. For example, using Au/Co/Au multilayer sandwich films probed by XMCD, it has been shown [12] that the perpendicular orientation of the total moment is a consequence of the fact that the orbital moment of Co films becomes highly anisotropic with decreasing film thickness. This orbital moment, perpendicular in thin Co films, is able to redirect the spin moment along the same perpendicular direction, overcoming the in-plane shape anisotropy due to the spin–spin dipole interaction. The driving force behind this alignment is the strong orbital anisotropy of the Co film and the resulting lowering of MCA energy.

In the early nineties [13,14], it was pointed out that the ground state expectation values $\langle L_z \rangle$ and $\langle S_z \rangle$ (of orbital and spin moments) could be obtained using sum rules for transitions from p$_{3/2}$ (L$_3$ edge) and p$_{1/2}$ (L$_2$ edge) core levels to $3d$ levels in $3d$ transition metals.

These sum rules can be expressed as

$$\frac{\int_{L_3} \sigma_m(E)\,dE + \int_{L_2} \sigma_m(E)\,dE}{\int_{L_3} \sigma_t(E)\,dE + \int_{L_2} \sigma_t(E)\,dE} = \frac{\langle L_z \rangle}{2N_h}, \tag{4.10}$$

and

$$\frac{\int_{L_3} \sigma_m(E)\,dE - 2\int_{L_2} \sigma_m(E)\,dE}{\int_{L_3} \sigma_t(E)\,dE + \int_{L_2} \sigma_t(E)\,dE} = \frac{\langle S_z \rangle + 7\langle T_z \rangle}{3N_h}. \tag{4.11}$$

These (many-body) sum rules include only the $2p \rightarrow 3d$ transitions and cannot account for effects such as $s - p - d$ mixing. Using local density theory based band calculations, Wu *et al.* [15], showed that the $\langle L_z \rangle$ sum rule was accurate to within 10%, if effects for band hybridization could be accounted for experimentally. They also argue that the magnetic dipole term, $\langle T_z \rangle$, can be important in non-cubic solids, contrary to the assumptions of Carra *et al.* [14]. O'Brien and Tonner [16] and others have explored the accuracy issues of the sum rules in some detail and concluded that they are qualitatively correct.

The linear counterpart of XMCD is X-ray magnetic linear dichroism (XMLD). Although XMLD predates the discovery of XMCD, its use in anisotropy measurements was demonstrated much later, for example, in an experiment using vicinal Co with varying step densities [17]. The interpretation of XMLD is considered less straightforward than XMCD, but a sum rule has been proposed which designates XMLD also as a probe of MCA [18].

4.1.7. *Mermin–Wagner theorem*

If one considers an array of spins, coupled by a Heisenberg-type exchange interaction, $J\mathbf{S}_i \cdot \mathbf{S}_j$, this system would exhibit rotational symmetry, i.e., those spins would be free to rotate rigidly in space without altering any physics with no preferred direction for the spins. The Mermin–Wagner (M-W) theorem is a rigorous result in statistical mechanics, which shows the absence of ferromagnetic or antiferromagnetic long range order in one- or two-dimensional isotropic spin **S** Heisenberg model with finite range exchange interactions *at any nonzero temperature*. According to the M-W theorem [19], if the Hamiltonian of a 2D spin system is invariant under a rigid rotation of all the spins, then the system cannot sustain long range (FM) order at finite temperature. The basic reason behind this is that the spin waves generated at finite temperature are so destructive that they kill ferromagnetic (or antiferromagnetic) long range order. Mathematically, this can be tied to a dispersion k^2 of the spin waves (for small k, i.e., long wavelengths) and the (logarithmic) divergence of an integral representing the number of excitations representing spin waves, as follows. The magnetization of a spin lattice can be written as

$$M = M_s - \frac{g\mu_B}{2V} \sum_\mathbf{k} n(\mathbf{k}). \tag{4.12}$$

The sum over $\mathbf{k}$ can be converted into an integral and in two-dimensions; this sum, related to the number of spin waves, diverges logarithmically as $k \to 0$.

So the question is how to reconcile the above with real experimental systems, such as thin metallic films, which exhibit long range magnetic order. The important clue here is that spin system has to be isotropic for the M-W theorem to hold. When there is anisotropy as in real systems, the theorem is inconclusive (the Heisenberg model, with nearest neighbor interactions, describes local spin–spin interactions). We know that lower symmetry is almost always accompanied by significant changes in anisotropy.

4.1.8. *Spin Hamiltonian*

As we have mentioned throughout this chapter, the spin–orbit interaction plays a central role in any discussion of magnetic anisotropy. The (local) orbital angular momentum couples with the (total) spin of a given atom to give rise to a local energy term. Now imagine a neighboring atom having a slightly different orbital angular momentum. When this atom couples to its own spin, pointing (say) in the same direction as the previous one, it will result in a slightly different local energy term. By minimizing the sum of such energy terms for atoms in a unit cell, an easy direction for the total moment can be found.

The "Spin Hamiltonian" is a perturbative treatment of the spin–orbit (S-O) interaction and the effects due to an external magnetic field. The perturbation is

$$\mathcal{H} = \xi \mathbf{L} \cdot \mathbf{S} + \beta \mathbf{H} \cdot (\mathbf{L} + 2\mathbf{S}), \tag{4.13}$$

where $\mathbf{H}$ is the external magnetic field, while $\mathbf{L}$ and $\mathbf{S}$ are the orbital and spin operators. The spin–orbit coupling constant, ξ, depends on the gradient of the potential. The perturbative term is calculated using the unperturbed, ground state multiplet.

It can be argued that such a perturbative treatment of the S-O-I is valid for $3d$ transition metals, where spin–orbit effects are not that large, compared to other competing interactions such as the crystal field. In this case, the wave functions can be approximated as a product of spin and orbital terms. However, one must keep in mind that for $5ds$, lanthanides and actinides (with correlated f electrons), this treatment may be not adequate and could lead to misleading results.

In the Spin Hamiltonian approach, only even orders of perturbation theory enter as long as time reversal (TR) symmetry is not broken. We note that the while the spin–orbit term does not break TR symmetry, an external magnetic field will do so resulting in a first order perturbation term which reads $\langle 0|\mathcal{H}|0\rangle = 2\beta \mathbf{H} \cdot \mathbf{S}$ for an orbitally nondegenerate ground state, $|0\rangle$. For such a state, the second-order correction term due to spin–orbit interaction is,

$$H_{SO}^{(2)} = \sum_{\alpha,\beta} K_{\alpha,\beta}^{(2)} S_\alpha S_\beta \tag{4.14}$$

with

$$K_{\alpha,\beta}^{(2)} = \lambda^2 \sum_{n \neq 0} \frac{\langle 0|L_\alpha|n\rangle \langle n|L_\beta|0\rangle}{E_0 - E_n}. \tag{4.15}$$

One can see that the asymmetries in the matrix elements $\langle 0|L_\alpha|n\rangle$, when they are nonzero for directions $\alpha = x, y, z$ will lead to the anisotropies at this level. For a cubic system, the above term will not depend on the direction of spin (hence symmetric in x, y, z), since the matrix $K^{(2)}$ is diagonal as well as isotropic. For such systems, a directional dependence in $K_{\alpha,\beta}$ will be seen at fourth order.

4.1.9.   Band theoretical treatments

Instead of directly using the "Spin Hamiltonian" to perturbatively calculate the spin–orbit interaction energy, MAE has also been calculated in terms of a sum of single particle eigenvalues, with and without the spin–orbit interaction, i.e.,

$$MAE = \sum [S - O - I]\epsilon_i - \sum [no - S - O - I]\epsilon_i. \tag{4.16}$$

In $3d$ metals, due to orbital quenching, this energy can be extremely small and calculating it accurately is by no means an easy task. Gay and Richter [20], in their pioneering work, moved beyond the above and used energy band states (from first principles) to estimate the S-O-I term perturbatively for monolayers of ferromagnetic Fe, Ni and V. First, they carried out spin-polarized band calculations, ignoring the spin–orbit term, to self-consistency and then the S-O-I was evaluated invoking a direction for the spin.

By using symmetry considerations (as in Equations (4.3) and (4.4)), the spin–orbit term was expressed as

$$E_{s.o.} = E_{s.o.}^{(0)} + E_z^2 \alpha_z^2 + E_{xz}^4 (\alpha_x^2 \alpha_z^2 + \alpha_y^2 \alpha_z^2) + E_{xy}^4 \alpha_x^2 \alpha_y^2 + O(\alpha^6), \quad (4.17)$$

where α_x, α_y, α_z are the direction cosines of the quantization direction of spin with respect to $\hat{x}$, $\hat{y}$ and $\hat{z}$ directions. The parameters could be determined from the spin–orbit energy calculations, when a specific direction was assigned to spin.

By following the anisotropy energy as a function of the spin quantization direction, it was determined that for a monolayer of Fe(001), the moments were oriented perpendicular to the plane of the film while for Ni, they were parallel to the surface at the monolayer thickness. The direction of the total moment is determined by the competition of the dipolar term and the spin–orbit term. For thicker layers of Fe, the dipolar term eventually wins out and the moments tend to be parallel to the plane for such films.

These calculations were quite tedious and required many thousands of k-points in the Irreducible Brillouin Zone. Later, Wang *et al.* [21] introduced a "state tracking" system to improve the convergence related problems. However, it must be mentioned here that in addition to numerical problems, LDA based wavefunctions used in evaluating the estimates do not include orbital polarization issues discussed above, simply because LDA based density functionals have no orbital specific information included.

4.1.10. Spin reorientation transitions in multilayers

Spin reorientation transitions demonstrate the competition among various anisotropy terms discussed so far and have now been observed in several multilayer systems. Naturally, these are of fundamental and practical interest. The reorientations occur when an appropriate parameter (such as the film thickness or temperature) is varied [22,23]. Here we consider two such examples, (a) revisiting a familiar multilayer system, fcc Fe/Cu/Fe trilayers [24], where film type and thickness drive the transitions and (b) a reorientation transition driven by temperature in ultra-thin Fe films on Gd(0001) [25] as illustrated in Figure 4.4. In the former experiment, fcc (γ) Fe films stabilized on Cu(001) were used as a template for the fabrication of the trilayer system. Below a thickness of ≈ 4 ML of fcc Fe, the easy axis for magnetization was found to be normal to the plane (with substrate temperature held at 300 K in an ultra high vacuum). The polar MOKE hysteresis loop in Figure 4.5(a) shows this well known result. The deposition of 2 ML of Cu on the above multilayer causes a reorientation of the magnetization into the plane of the film, as seen from Figure 4.5(b) for the longitudinal Kerr signal (with no polar Kerr effect). When an additional 3 ML of Fe were deposited on this system, the magnetization reoriented itself in the normal direction.

In the second experiment, Fe films grown on Gd(0001) exhibit a magnetic, in-plane to out-of-plane spin reorientation transition with increasing temperature. Such transitions probe the temperature dependence of the anisotropy constants, $K_2(T)$ and $K_4(T)$. According to this study, at low temperatures, the Fe moment remains in-plane, coupled antiferromagnetically to Gd. At intermediate temperatures, the Fe/Gd surface undergoes

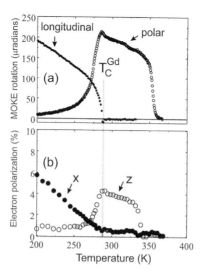

Figure 4.4.   Fe/Gd spin reorientation transition driven by temperature. Reproduced with the permission of the American Physical Society from Ref. [25].

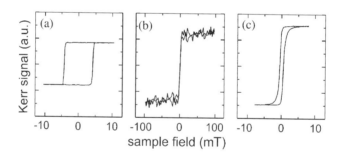

Figure 4.5.   MOKE hysteresis loops. (a) and (c) refer to polar and (b) to longitudinal MOKE hysteresis loops. (a) 3 ML of Fe, (b) 2 ML Cu/3 ML Fe, (c) 3 ML Fe/2 ML Cu/3 ML Fe. Reproduced the with permission of Elsevier from Ref. [24].

a continuous rotation to a canted direction. From polar and longitudinal MOKE measurements, in addition to spin-polarized secondary emission spectroscopy shown in Figure 4.4 and susceptibility measurements, it was argued that the reorientation transition occurs in two steps. First is the continuous rotation to a canted direction mentioned above while the second is a thermally irreversible rotation from this canted direction to a direction perpendicular to the plane of the film.

This experiment on trilayers (see Figures 4.5 and 4.6) attempts to make the point that magnetic anisotropy is the driving force behind the observed alignment of magnetization, not the modified RKKY type interaction which is related to the GMR effect.

Another interesting issue related to spin reorientation transitions has to do with the critical region (in temperature or thickness) where the effective anisotropy vanishes and the magnetization direction changes (between parallel and perpendicular) as evident from Figure 4.7 in Fe/Ag(100) (Ref. [23]). If magnetic anisotropy is responsible for the long

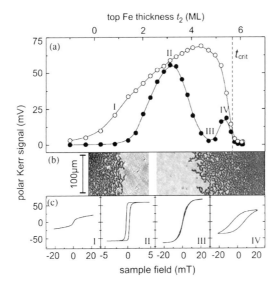

Figure 4.6. fccFe/Cu trilayer system. The variation of MOKE signal with respect to increasing the top layer thickness is shown. Above a critical thickness of t_{cr}, there is no perpendicular magnetization. Reproduced with the permission from Elsevier from Ref. [24].

range order (LRO) seen in magnetic films, does the M-W theorem, which precludes long range order in 2-dimensional Heisenberg systems at non-zero temperature, hold in this critical region? Although early theoretical studies suggested a lack of LRO, SMOKE measurements were able to conclude that there was some non-zero (but suppressed) magnetization. Apparently, there is nonzero magnetization with a complex magnetic (domain) structure referred to as stripes. A "pseudo-gap", where the magnetic domains are favored, is found in this critical region leading to some fascinating physics [23]. Yafet and Gyorgy [26] have shown that in a 2-dimensional Heisenberg system with uniaxial anisotropy, stripe domains are favored over a single domain. Large domains (of the order of a micron) could be formed with the size of the critical region depending on the strength of the effective anisotropy. Kashuba and Pokrovsky [27] have demonstrated an analogy with a liquid crystal that has orientational order but no LRO, while Allenpasch and Bischof [28], through their SEMPA analysis, have shown that the single domain disintegrates into (randomly distributed) multiple stripe domains with orientational order in the transition region.

4.2. Exchange-bias due to exchange anisotropy

The anisotropies discussed above are either due to spin–orbit coupling or magnetic dipolar fields arising from size effects. A somewhat different anisotropy, resulting from 'frustrated' exchange interactions found at AFM/FM interfaces, has found widespread use in applications such as read heads of (magnetic data storage) hard drives. In such systems, uniaxial anisotropy is induced in a ferromagnetic layer by adjacent antiferromagnetic layers. The phrase exchange anisotropy was coined to describe this phenomenon, which has

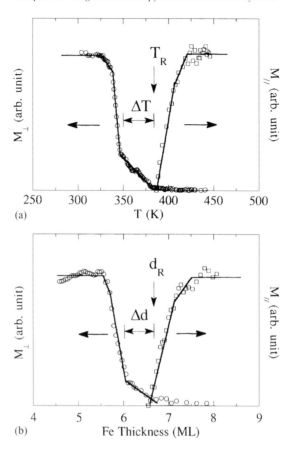

Figure 4.7. Critical regions (in temperature and film thickness) in the spin-reorientation transition of Fe/Ag(100) discussed in the text. Reproduced with the permission of the American Physical Society from Ref. [23].

defied a straightforward quantitative analysis up to now. Before discussing its relevance to magnetic data storage, we will cover the basics of exchange anisotropy and its complexity which will in turn be useful in understanding the theoretical difficulties.

In soft ferromagnetic materials, it is possible to switch the direction of magnetization by applying an external magnetic field. As is well known, a hysteresis loop, that shows a limited symmetry (centering about $h = 0$) as the applied field $h_{app} \to \pm h$, is seen in such materials. When antiferromagnetic layers are used to pin the magnetization of the ferromagnet, this symmetry along the field axis is broken. Exchange bias effect is usually associated with well defined ferromagnetic/antiferromagnetic interfaces, with frustrated spin–spin interactions. This effect was first observed in the 1950s in Co particles embedded in their antiferromagnetic oxide [5,6]. Apparently, this discovery was, as some significant discoveries are, a serendipitous one. Meikljohn and Bean [5,6] were trying to demonstrate the expected coercivity of Co particles, when they noticed the effect of its oxide (CoO), in the form of a shift, on the hysteresis loop. When the oxide was absent (or removed) the hysteresis loop was symmetrical.

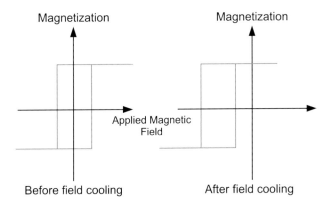

Figure 4.8.   A simple illustration of the exchange-bias effect.

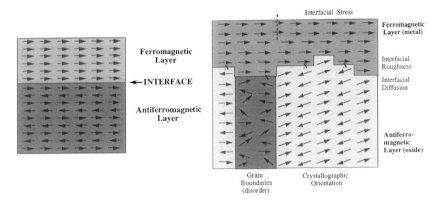

Figure 4.9.   A schematic of possible interfacial spin arrangements in an exchange-bias setup showing a perfect interface (left) and an interface which is more realistic (right). Reproduced with the permission of Elsevier [29].

Exchange-bias, as the name loosely suggests, describes a shift (or a bias) H_E of a magnetic hysteresis loop along the field axis from its $h = 0$ centered curve under the reversal of an applied field h (see Figure 4.8). It is now recognized that the interfacial spin configuration plays a crucial in the exchange-bias effect (Figure 4.9). A highly simplified view of this phenomenon can be described as follows: The system under consideration is a (soft) ferromagnetic layer on top of an antiferromagnetic material, where the Néel temperature of the antiferromagnet is assumed to be less than the Curie temperature of the soft ferromagnet (i.e., $T_N \leqslant T_C$). The magnetization of the soft FM can be rotated using an applied field and its hysteresis would show symmetry about h_{app}, as it is reversed, in the absence of the antiferromagnet.

However, when the antiferromagnet is present and cooled below its Néel temperature as indicated in Figure 4.8, its uncompensated magnetic moments couple to the FM layer and an exchange-bias is locked in. A compensated AFM interface is where the net spin averaged over a microscopic area turns out to be zero, while at an uncompensated region this average is nonzero (see Figure 4.9). Such interfaces and related bias effects have

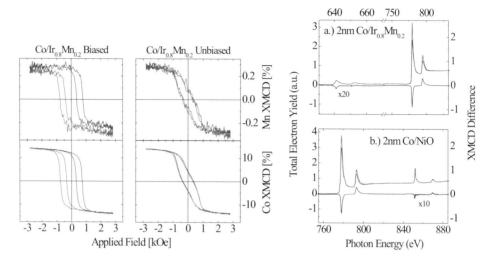

Figure 4.10.   Left: Illustration of element specific XMCD hysteresis loops and exchange-bias effect from Ref. [32]. Magnetization is measured on Co and Mn in a Co(2 nm)/$Ir_{20}Mn_{80}$(50 nm) multilayer. The hysteresis loops were acquired with the bias field (horizontal loop shift) either parallel or antiparallel to the propagation direction of the incident X-rays. Right: XMCD spectra for Co/$Ir_{20}Mn_{80}$ and Co/NiO for parallel (red) and antiparallel (blue) alignment of the external field and X-ray helicity as well as their difference (black). Uncompensated and rotatable spins are responsible for the observed Mn and Ni dichroism. Reproduced with the permission of the American Physical Society [32].

been observed in FM/AF systems such as NiFe/FeMn, NiFe/CoO, Fe/FeF_2 etc., where the Curie temperature of the FM is higher than the Néel temperature of the AFM. An element specific exchange-bias effect where the magnetization is measured on Co and Mn in a Co(2 nm)/$Ir_{20}Mn_{80}$(50 nm) multilayer is shown in Figure 4.10 (Ref. [32]). The exchange coupling at the interface would either be favorable (helpful) to the magnetization of the ferromagnet or would hinder (resist) when an applied field is used to rotate the orientation of magnetization. Clearly, there is a symmetry breaking here. The need for an AFM (as opposed to a FM) as the supporting layer may be tied to the ease with which the above rotation can be carried out (i.e., with ferromagnetic pinning, it may be harder to change magnetization). Figure 4.11 shows various possible stages of the spin arrangements as the applied field is varied [30].

Early on, a simple model, starting from the Stoner–Wohlfarth free energy expression, was introduced by Meikljohn and Bean (MB) to explain the exchange-bias effect [5,6]. This model is capable of giving only a qualitative description of the exchange-bias effect, but is regarded as a reasonable starting point. Even under the assumption of a perfect interface, there is no general quantitative agreement with experiment. However, the following resulting equation points to the importance of nonzero magnetizations at both sides of the "interface", ferromagnetic as well as antiferromagnetic:

$$\mu_0 H_e = -\frac{J I_{AF} I_{FM}}{M_{FM} t_{FM}}. \qquad (4.18)$$

Here I_{AF} and I_{FM} refer to interface magnetizations of the antiferromagnet and the ferromagnet respectively. Note that the MB approach is based on an idealized (perfectly flat)

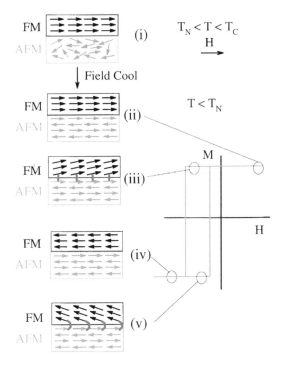

Figure 4.11.   A schematic illustration of spin configurations of a FM/AFM bilayer at different stages (i)–(v) of an exchange-biased hysteresis loop. Note that this is a simple cartoon to illustrate the effects of coupling and they are not necessarily actual portraits of a realistic situation. Reproduced with the permission of Elsevier from Ref. [30].

interface, consisting of a coherently rotating ferromagnet and an antiferromagnet with strong uniaxial anisotropy.

4.2.1.   Exchange anisotropy with insulating AFM films

Many of the reported AFM films with insulating behavior (used in exchange biasing) are monoxides such as NiO, CoO or $Ni_xCo_{(1-x)}O$ with the notable exception of FeF$_2$. These have the NaCl-like cubic structure and a slight distortion below the Néel temperature (T_N) with parallel spins on (111) planes, coupled antiferromagnetically to adjacent planes, as shown in Figure 4.12. The Néel temperatures for bulk NiO and CoO are 525 K and 293 K (respectively) with a linear variation for $Ni_xCo_{(1-x)}O$ with x. Two anisotropy constants associated with in-(111)-plane and out-of-(111)-plane rotations can be identified with CoO showing noticeably larger anisotropies compared to NiO. (See Ref. [29] for more details.) When one of these (111) planes forms an interface with a ferromagnetic material, the spin structure at the interface region is said to be uncompensated. Uncompensated interfacial spins associated with various systems such as, CoO/MgO, Co/NiO, have been observed.

In particular, X-ray magnetic circular dichroism (XMCD) experiments showed that the existence of uncompensated spins is not sufficient for exchange-bias, if they get locked

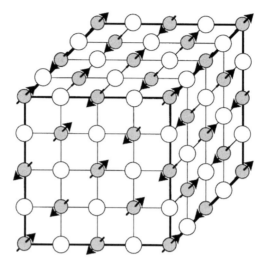

Figure 4.12. Antiferromagnetic ordering in the monoxides, such as NiO in the rocksalt structure, used in exchange biasing mentioned in the text. The neighboring (111) planes which contain the magnetic cation (Ni) are aligned antiferromagnetically with each other.

into the FM and rotate with those (FM) spins under an applied field. Such a scenario will not yield an exchange-bias. Some of the lessons learned from the oxide work are; (a) uncompensated spins at the interface determine the loop shift H_E, (b) the coercive field H_C increases with lower AFM anisotropy, (c) in polycrystalline sample, repeated cycling of the applied field resulted in lowering of both H_E and H_C (now known as "training").

Magnetic anisotropy in NiO films has been studied rather extensively, especially with regard to exchange-bias related phenomena. In one such study, NiO films were exchange-biased using ferromagnetic $Co_{84}Fe_{16}$ layers [31]. Magnetic linear dichroism using soft X-rays showed exchange-bias induced magnetic anisotropy in both (111) textured and untextured NiO films with the Ni moment axis being parallel to the exchange-bias field direction. This is claimed to be one of the first observations of a key step in exchange-biasing, namely, the repopulation of the AFM domains whose axis is closest to the exchange-bias field direction.

4.2.2. Exchange anisotropy with metallic AFM films

Most of the metallic AFM films used in exchange-biasing are Mn alloys. In most applications these are preferred due to higher Néel temperatures and larger anisotropies. However, the basic understanding obtained from AFM oxide based systems have not undergone any significant changes due to results from the metallic AFM systems. For example, in a notable, recent experiment using XMCD, it has been claimed that only a small percentage of the interfacial spins are actually pinned to the AF, when an exchange-bias effect occurs (see Figure 4.13) [32]. These pinned, uncompensated spins do not rotate in an external magnetic field and they are tightly locked into the AFM lattice. As observed with the AFM oxide films, if this pinned (spin) percentage is only a small fraction of a monolayer, it would explain the small exchange bias field shifts H_E seen in the experi-

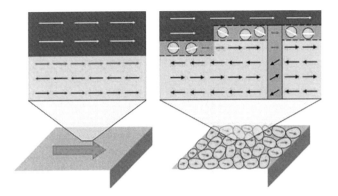

Figure 4.13. Schematic view of the pinned and uncompensated spins associated with the exchange-bias effect as suggested in Ref. [32]. An ideal antiferromagnet (in green) is shown on the left where all the interfacial spins are pinned. On the right, a more realistic, rough interface is shown with most spins at the interface not pinned and hence able to rotate with the ferromagnet. The basic idea here is that only a small percentage of the interfacial spins are pinned (red arrows) and hence responsible for the observed exchange-bias. For interpretation of the references to colour in this figure legend, the reader is referred to the web version of this chapter.

ments. The large quantitative disagreements with the MB model can also be understood based on the above observations. This is exactly the argument presented in Ref. [32]; their experimental findings are briefly summarized below.

In the sandwich structure $Co/Ir_{20}Mn_{80}$, both the Co spins in the FM and the Mn spins in the AFM exhibited identical coercivities, bias fields and revealed similar shapes in hysteresis loops. In addition, the Mn loops of the field deposited sample showed a small vertical shift, which was absent in the zero-field deposited sample. This vertical shift was directly linked to the pinned, uncompensated spins and was used to estimate that a small fraction (about 7%) of the total uncompensated moments (0.56 ± 0.14 ML) is pinned. A majority of the uncompensated spins appeared to rotate with the ferromagnet.

The amount of pinned, interfacial magnetization is correlated to the observed macroscopic bias fields following the MB model. The bias field H_E can be written as a function of the unidirectional magnetic interface energy σ as

$$H_E = \frac{\sigma}{M_{FM}t_{FM}} = -\frac{J I_{AF} I_{FM}}{M_{FM}t_{FM}}. \qquad (4.19)$$

This model can be applied for ideal AFM/FM sandwich structures but fails for non-ideal systems, overestimating the interfacial coupling strength by orders of magnitude. However, taking into account the small fraction of pinned magnetization at the interface, the ideal coupling J and interfacial energy density σ can be replaced by J_{eff} and σ_{eff} respectively, to yield reasonable bias fields.

4.2.3. Applications: Exchange-biasing in spin-valves/sensors

In magnetic recording, the write element records signatures (bits of information) on a thin film surface coated on a hard disk. These bits are either "0" or "1" (depending on the direction of magnetization) and have to be detected with a read element that senses the magnetically recorded information from a narrow region. As discussed earlier (in

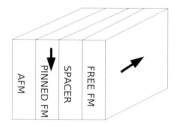

Figure 4.14.   Schematic illustration of the sandwich structure in a spin-valve sensor.

Chapter 1), this sensing is done utilizing the GMR (or AMR) effect, i.e., the dependence of the resistance on the alignment of spins on either side of a nonmagnetic spacer. Even before the discovery of the GMR effect, exchange-biasing was used in magnetic recording read heads utilizing the AMR effect. However, the percentage change in resistivity is much higher with GMR sensors and they are sensitive to smaller changes in magnetic field variations when compared to AMR based sensors. The read head (spin-valve) is suspended in air above the hard disk in the region of interest (i.e., where the data to be read are located). The size of such a read head has to match the bit-size, since otherwise information from other bits will affect the reading, giving rise to incorrect interpretations of the written data. The stray fields that identify the changing of a bit value (say) from "0" to "1" along the magnetic strip causes a rotation of the moment in a soft ferromagnetic layer (i.e., a free FM or 'sensing' layer) in the read head; this forms a part of a sandwich structure which has to have another FM layer, separated via a spacer. The second FM layer is referred to as the reference layer and carries the pinned magnetic moment. Pinning is achieved in the sandwich via exchange-biasing with an AFM layer as discussed in this chapter. Due to uniaxial anisotropy resulting from exchange-biasing, the magnetization of the reference layer is essentially fixed, provided that the bias field H_E is higher than the stray fields being measured, while the sensing layer FM moments are free to rotate. (See the sandwich structure depicted in Figure 4.14.) This rotation gives rise to a change in resistivity that can be measured.

However, there are several (engineering) issues that have to be taken into account when optimizing such devices used today in advanced recording heads. There could be substantial magnetic coupling between the sensing and reference layers due to demagnetizing fields generated from the dipoles at the edges, which will favor antiparallel coupling. In addition, roughness related dipolar fields can favor a different coupling. Finally, the currents that are used to sense the change in resistivity themselves generate magnetic fields. In present-day spin-valve read heads, all these effects have to be carefully balanced. For hard disks having areal densities of about ~100 Gb/in², a typical read sensor element has a width of about ~100 nm and a height of about ~70 nm [33]. Such tiny dimensions naturally lead to significant shape anisotropies which must be taken into account when trying to optimize these spin-valves.

In summary, the phenomenon of exchange-bias has attracted a significant amount of interest due to numerous, possible applications. Exchange-bias depends critically on the atomic scale arrangement of the spins at or near an interface. In addition to the AFM/FM materials used, the horizontal shift of the field, H_E, depends on both intrinsic (such as anisotropy) and extrinsic (such as interface roughness, training) effects. Consequently, it has been a challenging topic and defied numerous theoretical attempts to obtain a quantitative understanding. The extrinsic effects make the task at hand harder, and improved layer quality and a better understanding of the atomic scale spin structure at the interface will be crucial for further developments in the field.

References

[1] B. Sinkovic, Private communication.

[2] D. Weller, Y. Wu, J. Stöhr, M.G. Samant, B.D. Hermsmeier, Phys. Rev. B **49**, 12888 (1994).

[3] J.H. van Vleck, Phys. Rev. B **52**, 1178 (1937).

[4] L. Néel, J. de Phys. et le Rad. **52**, 225 (1954);
L. Néel, J. Phys. Radium **15**, 227 (1954).

[5] W.H. Meiklejohn, C.P. Bean, Phys. Rev. **102**, 1413 (1956).

[6] W.H. Meiklejohn, C.P. Bean, Phys. Rev. **105**, 904 (1957).

[7] W.H. Meiklejohn, J. Appl. Phys. **33**, 1328 (1962).

[8] F. Keffer, Phys. Rev. **87**, 608 (1952);
Y. Shapira, S. Foner, Phys. Rev. B **1**, 3083 (1970).

[9] P.F. Carcia, A.D. Meinhaldt, A. Suna, Appl. Phys. Lett. **47**, 178 (1985).

[10] G.H.O. Daalderop, P.J. Kelly, F.J.A. den Broeder, Phys. Rev. Lett. **68**, 682 (1992).

[11] F.J.A. den Broeder, E. Janssen, W. Hoving, W.B. Zepper, IEEE Trans. Magn. **28**, 2760 (1992).

[12] D. Weller, J. Stöhr, R. Nakajima, A. Carl, M.G. Samant, C. Chappert, R. Mégy, P. Beauvillain, P. Veillet, G.A. Held, Phys. Rev. Lett. **75**, 3752 (1995).

[13] B.T. Thole, et al., Phys. Rev. Lett. **68**, 1943 (1992).

[14] P. Carra, et al., Phys. Rev. Lett. **70**, 694 (1993).

[15] R. Wu, D. Wang, A.J. Freeman, Phys. Rev. Lett. **71**, 3581 (1993);
R. Wu, A.J. Freeman, J. Appl. Phys. **79**, 6209 (1996).

[16] W.L. O'Brien, B.P. Tonner, Phys. Rev. B **17**, 12672 (1994).

[17] S.H. Dhesi, G. van der Laan, E. Dudzik, A.B. Shick, Phys. Rev. Lett. **87**, 067201 (2001).

[18] G. van der Laan, Phys. Rev. Lett. **82**, 640 (1999).

[19] N.D. Mermin, H. Wagner, Phys. Rev. Lett. **17**, 1133 (1966).

[20] J.G. Gay, R. Richter, Phys. Rev. Lett. **56**, 2728 (1986).

[21] D.-S. Wang, R. Wu, A.J. Freeman, Phys. Rev. Lett. **70**, 869 (1993).

[22] D.P. Pappas, K.-P. Kamper, H. Hopster, Phys. Rev. Lett. **64**, 3179 (1990).

[23] Z.Q. Qiu, J. Pearson, S.D. Bader, Phys. Rev. Lett. **70**, 1006 (1993).

[24] A. Enders, D. Repetto, T.Y. Lee, K. Kern, J. Magn. Magn. Mater. **272–276**, e959 (2004).

[25] C.S. Arnold, D.P. Pappas, A.P. Popov, Phys. Rev. Lett. **83**, 3305 (1999).

[26] Y. Yafet, E.M. Georgy, Phys. Rev. B **38**, 9145 (1988).

[27] A. Kashuba, V.L. Pokrovsky, Phys. Rev. B **48**, 10335 (1993).

[28] R. Allenpasch, A. Bischof, Phys. Rev. Lett. **69**, 3385 (1992).

[29] A.E. Berkowitz, K. Takano, J. Magn. Magn. Mater. **200**, 552 (1999).

[30] J. Nogués, I.K. Schuller, J. Magn. Magn. Mater. **192**, 203 (1999).

[31] W. Zhu, L. Seve, R. Sears, B. Sinkovic, S.S.P. Parkin, Phys. Rev. Lett. **86**, 5389 (2001).

[32] H. Ohldag, et al., Phys. Rev. Lett. **91**, 017203 (2003).

[33] S.S.P. Parkin, X. Jian, C. Kaiser, A. Panchula, K. Roche, M. Samant, Proceedings of the IEEE **91**, 661 (2003).

Chapter 5

Probing Layered Systems: A Brief Guide to Experimental Techniques

This chapter briefly describes several experimental techniques, which have been regularly referred to in this volume, that are used to characterize multilayer systems. Some of these techniques are not exclusively restricted to multilayers and this is by no means an exhaustive or a detailed list but the goal is to provide the non-specialist reader with some background information on the relevant experimental methods. One thing to note here is that although they have been in use for some time, some of the techniques described here have become quite sophisticated over the years.

5.1. SMOKE (Surface Magneto-Optic Kerr Effect)

Although the Kerr effect has been known for quite some time, the modern application of its surface magneto-optic version was pioneered by Bader and his colleagues at Argonne National Lab (Refs. [1–3]). Linearly polarized light can change its polarization (and become elliptically polarized) upon reflection from a magnetic surface (or in the presence of a magnetic field). Depending on the direction of magnetization, different forms of polarizations of the reflected beam can occur. There are 3 Kerr configurations that are of significance [2,3]. They are: (1) polar, (2) longitudinal, (3) transverse. In the polar Kerr effect, the magnetization is perpendicular to the plane of the film, while for the longitudinal one, the magnetization is in the plane of the film and in the scattering plane of light. In the transverse effect, the magnetization is in the plane but perpendicular to the plane of incident light. In the latter, there is no change in the polarization of light since there is no component of the magnetization in the direction of propagation of light. The polar Kerr effect, in general, gives rise to larger polarization changes when compared with the longitudinal one. An important advantage in SMOKE appears to be associated with the fact that a relatively simple, *in situ* analysis of the polarized beam can be carried out external to the UHV chamber where the sample is located. In addition, it is inexpensive and simple to setup. Due to these reasons, it has become *the technique of choice* for basic characterization of magnetic films. (See Figures 5.1 and 5.2 for a schematic SMOKE setup.)

The microscopic origin of this effect when light is reflected from a magnetic surface is the spin–orbit interaction. The direction of magnetization is determined by the spin–orbit interaction and this, in turn, will interact with the incident photon. A macroscopic description, based on Maxwell's equations, can be given as follows: Linearly polarized light

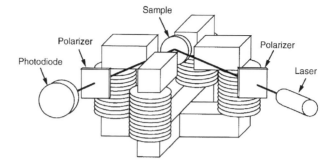

Figure 5.1.   Schematic SMOKE setup. The top part illustrates the change in polarization of light when reflected from a magnetic material. Reproduced with permission of Elsevier from Ref. [3].

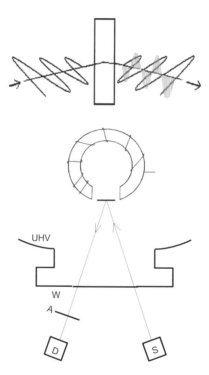

Figure 5.2.   SMOKE setup that allows the measurement of longitudinal and polar hysteresis loops. Reproduced with permission of Elsevier from Ref. [3].

can be decomposed into left-handed and right-handed circularly polarized modes with equal amplitudes. However, when reflected, these modes can have different reflection coefficients and, therefore when combined, will give rise to a Kerr rotation. (To be more precise, the magnetic field is an axial (or pseudo) vector whose transformation properties are different from true vectors, and can recognize right and left circular polarization.) Microscopically, one can imagine this effect as being due to lifting the degeneracy associ-

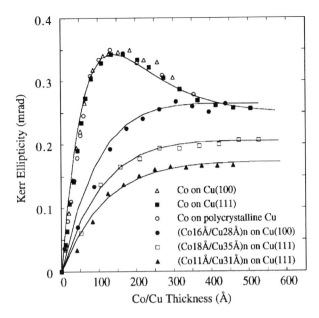

Figure 5.3. Variations of Co/Cu Kerr ellipticity as a function of the total film thickness. Reproduced with permission of Elsevier from Ref. [3].

ated with the magnetic quantum number m (due to breaking the time reversal symmetry). This in turn implies that right circularly polarized light may be able to excite electrons to some empty level with an appropriate m value but the left-circularly polarized light may not (since the corresponding $-m$ state may be occupied). To be more accurate, when dealing with partially filled states the situation is more complicated. Also, a more suitable quantum number is the total angular momentum (and its components) in the presence of a strong magnetic field or spin–orbit interaction. (This is discussed further in the chapter on first principles theory.) However, the asymmetry associated with the above situation can be detected and analyzed to provide information about surface magnetic anisotropy or quenching of orbital moments.

Significant insights into changes in the magnetic states have been obtained from *in situ* SMOKE experiments conducted on thin films. As an example, a reversible switching of the easy axis of magnetization for Ni/Cu(001) has been reported using MOKE [10]. Another example is when growing magnetic layers on a nonmagnetic substrate (such as Co on Cu as shown in Figure 5.3), changes in magnetic anisotropy could be monitored (*in situ*) as a function of (time and) layer thickness. Usually SMOKE is carried out with incident light having a fixed wavelength (say corresponding to red in the visible region of the electromagnetic spectrum). However, since the Kerr rotation is associated with some intricate details of the electronic/magnetic structure, its magnitude is a complex function of the wavelength of the incident light used. There are such MOKE spectroscopic measurements, monitoring the magnitude of the Kerr rotation as a function of wavelength, and these experiments yield more detailed insight into the electronic structure of the sample.

5.2.   AES (Auger Electron Spectroscopy)

Auger spectroscopy is a useful tool for detecting the chemical nature of a given surface during film growth. When an atom is subjected to an energetic electron beam or X-rays, core electrons get ejected and holes are created inside the core shells. This could be regarded as the first necessary step. When these core-holes get filled by electrons from higher energy levels (i.e., when the core-hole decays), the net energy change can go into photons or to other (outer) electrons, generally referred to as Auger electrons. The energy of these Auger electrons depend on the atom under study, rather than the incident, energetic (primary) electrons. The Auger electrons, when properly identified and separated (for example, using a modulation technique [6]), provide a unique, chemical signature for the atom in question. The Auger process usually dominates over photon emission for a low energetic incident electron beam (<1000 eV) and the emitted Auger electrons come from the surface region, since low energy electrons have small penetration depths in solids. Due to its element selective nature and the surface sensitivity, it has become the leading technique for surface elemental analysis. It must also be noted that the Auger emission process involves two electrons, of relatively low binding energies. The transition begins with a missing core electron and ends with two missing electrons in the higher lying levels as indicated in Figure 5.4. If one or more of these electrons originate from the valence levels, the spectra will be sensitive to the local environment of the atom being probed. Further, the spectra will carry information about the correlations associated with the two electrons.

Chemical and surface sensitivity of AES is similar to X-ray Photoemission Spectroscopy (XPS) but is simpler in the sense that it does not require a monoenergetic beam as XPS does (only a simple electron-gun is needed). Thus AES is used more often compared to XPS in chemical as well as surface probes. A careful analysis of the Auger spectrum can yield information about the growth of the surface layers, especially at the initial stages [7]. There are excellent review articles on this topic (see, for example, Refs. [8,9]).

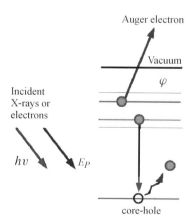

Figure 5.4.   Details of Auger Spectroscopy: Incident electrons create core holes which are responsible for the emitted secondary electrons.

5.3. FMR (Ferromagnetic Resonance)

As is well known, a magnetic moment **M**, placed in a magnetic field $\mathbf{H}_{eff}$ will undergo precession according to

$$\frac{d\mathbf{M}}{dt} = -\gamma \mathbf{M} \times \mathbf{H}_{eff}. \tag{5.1}$$

Usually, there will be some form of damping that will be responsible for the decay of the precession and eventually, the moment will align along the easy direction of magnetization of the medium. Ferromagnetic resonance is said to occur when the frequency of precession is identical to the frequency of an electromagnetic wave in which the medium is placed. This resonance was first observed by Griffiths [11] in 1946 and has now become a part of the arsenal used to characterize ultrathin films [12]. FMR is based on finding peaks (i.e., resonances) in the microwave absorption (or transmission minimum) and is a well established technique for measuring the magnetic anisotropy. The resonant frequency yields a direct measure of the local, effective field H_{eff}, while the width of the resonance provides a measure of the so-called Gilbert damping constant α. The effective local field is due to local anisotropy, applied field, exchange effects. In trilayer structures, two modes (optical and acoustic) are observed due to coupling of the ferromagnetic layers. The optical mode provides a straightforward measure of the interlayer coupling [13].

5.4. STM (Scanning Tunneling Microscopy)

Gerd Binnig and Heinrich Rohrer are credited with the invention of this elegant technique and were awarded the Nobel prize (in Physics) in 1986 for the development of the STM [14]. The STM is non-optical but can be used to obtain a direct image of a given region of a metallic surface. Unlike diffraction techniques, which depend on some form of long range order as well as an arduous analysis of the diffraction pattern, the STM is a local probe that does not require such order and yields a direct image. It is based on quantum mechanical tunneling of electrons from a tip (of the probe) to the surface that is being probed (or vice versa) as shown in Figure 5.5. When a voltage is applied between the tip (a sharp stylus made of metals such as Pt-Ir) and a metallic (sample) surface, a tunneling current flows between the two, provided that the tip is within a few angstroms of the surface. This is because of having a nonvanishing (but exponentially decaying) electron wavefunction (hence density of states) in the potential barrier represented by the region between the tip and the surface. When these exponential tails from the tip and the surface overlap, with an applied field, a tunneling current can be generated.

By probing a suitable region holding (say) the current constant, the surface morphology can be mapped directly with high resolution better than its newer cousin, Atomic Force Microscopy. It can obtain images at an atomic scale, up to resolutions of the order of 2 Å, and hence has the capability to map out atomic structure or even manipulate atoms. The tip of the microscope could be quite narrow and sharp (as narrow as a single atom) which is responsible for the high resolution. The exponential decay (with distance) of the current that tunnels between tip and surface in an STM ensures that the (monotonic) current is due to the front-most atoms of the tip and surface.

The STM is mostly used in post-growth situations and hence is not a technique that is used to monitor the (*in situ*) growth of layers. It can be used as a complementary method

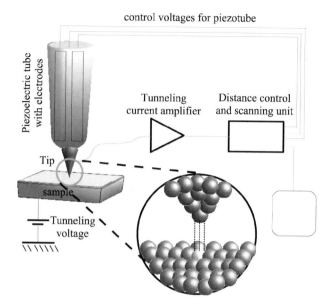

control voltages for piezotube

Tunneling
current amplifier

Distance control
and scanning unit

Piezoelectric tube
with electrodes

Tip

sample

Tunneling
voltage

Figure 5.5.   STM setup.

to diffraction studies, although here there is no need for periodicity of an underlying structure. The STM can also be used to do local spectroscopy of occupied and unoccupied states by measuring I/V curves at a fixed surface location. However, note that the STM cannot be used to probe semiconducting or insulating surfaces since these do not have electronic states that can tunnel to the tip.

5.5.   AFM (Atomic Force Microscopy)

The Atomic Force Microscopy (AFM) measures the atomic force between a tip and surface, and matured a decade or so later than its older cousin STM. The reason for this lag is that the surface forces acting on a tip are much more complicated that the tunneling current encountered in the STM. In an AFM, the total force consists of contributions from long and short range interactions, and may be nonmonotonic. This makes the feedback signal less tractable. In vacuum, the forces are attractive van der Waals type while closer to a surface, the forces associated with chemical bonding are attractive at first but become repulsive at sub-nanometer scale. A cantilever, which is the central operational piece of the technique, as shown in Figure 5.6, is used as a spring (with spring constant k) in order to detect the deflection (F/k) due to the force F exerted by the surface. The AFM is used more than the STM to estimate surface roughness, which is quite useful when dealing with thin films and their growth. Also note that since it measures a force, the AFM can be used for various types of surfaces (unlike the STM), such as semiconductors, metals, and insulating oxides. A landmark achievement of this technique was the first AFM image of the Si(111)-7 $\times$ 7 reconstruction. A comprehensive review of the evolution of AFM and recent developments are given in Ref. [15].

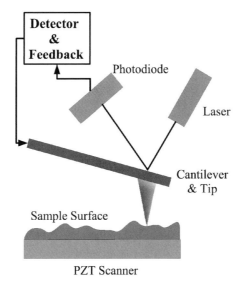

Figure 5.6.   Atomic Force Microscope.

5.6. Neutron diffraction

Neutron diffraction [16] has become an extremely useful experimental method for extracting information about spin configurations, such as ferromagnetic or antiferromagnetic ones. Neutrons are almost as heavy as protons without any charge but carry a magnetic moment (nuclear spin 1/2). The absence of charge and small velocities of the thermal neutrons makes them an attractive probe of condensed matter systems. The interaction of neutrons with condensed matter is manifestly two-fold: firstly, they interact with nuclei and secondly with electrons in the partially filled shells. The interaction with nuclei is isotropic since the size of nuclei is much smaller than the thermal neutron wavelengths (typically 1 Å). The microscopic magnetization is a vector quantity and hence non-isotropic. By analyzing the scattering of neutrons by electron spins (in the partially filled states only; core electrons do not contribute to magnetic scattering) information about magnetic order can be obtained [17].

Also, when combined with X-ray measurements, exact electron densities can be mapped out. In addition, neutron diffraction has played an important role in identifying structural phase transitions. Wide-angle neutron diffraction can provide useful information about domain sizes as well as ordering temperatures of magnetic systems. A drawback here is that neutron facilities are not that common and hence not readily available for the users.

5.7. Mössbauer spectroscopy

It had been long recognized that the γ-rays emitted when radioactive nuclei in excited states decay might excite other stable nuclei of the same isotope. However, when and if

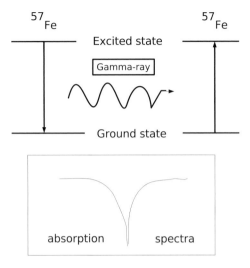

Figure 5.7.   Mössbauer principle.

the nucleus recoils, then the energy of the emitted γ-ray becomes less than the energy difference between two nuclear levels of interest and hence cannot excite another stable nucleus of the same isotope. For absorption too, it will be necessary to provide γ-rays with higher energies than the above energy difference. Hence for free nuclei, resonant absorption will be absent due to recoiling. However, in a solid, recoil-less absorption or emission is possible, with the energy of the radiation exactly matching the difference between the two nuclear levels (see Figure 5.7). In 1957, Mössbauer discovered that a nucleus in a solid can sometimes absorb or emit γ-rays without recoil. The probability of such a recoil-free absorption depends crucially on the energy of the radiation, and hence this effect is limited to certain isotopes with low lying excitations. The Nobel prize winning Mössbauer effect has now been detected in over one hundred isotopes, i.e., there are many nuclei of atoms in condensed matter that are capable of recoil-free absorption or emission of γ-rays.

The sharpness of the Mössbauer line and the ability to determine the energy positions of the emitted γ ray from a source relative to an absorber are two of the most important features of this method. The Mössbauer effect has been used to detect the Hyperfine interactions in solids providing a wealth of information regarding spin densities of electrons at the nucleus. For example, the isomer shift is a measure of s electron density (that can penetrate the electron cloud to reach the nucleus of an atom) interacting with the Coulomb interaction of the nuclear charge distribution over a finite radius. This interaction shifts the ground and excited state levels of the nuclei of the source and the absorber and can be detected by applying an appropriate Doppler velocity to either the source or the absorber. The isomer shift provides extremely useful information about the chemical environment and hence sometimes referred to as "the chemical shift". Nuclear Zeeman effect can also be detected using the Mössbauer effect. There many excellent review articles and books written about this method and Ref. [18] is one of them.

For studies of magnetic materials, the following experimental fact has been quite useful: ^{57}Fe isotope exhibits Mössbauer effect when excited with γ-rays from ^{57}Co. For such studies, the electrons emitted when the excited nucleus returns to its ground state constitute the spectra and this technique is referred to as Conversion Electron Möss-bauer Spectroscopy (CESM). When information about the local structure in a thin film is needed, a coating of ^{57}Fe is deposited at that location and CESM will yield information related to the local chemical as well as magnetic states (and environment) in the region of interest.

5.8. LEED (Low Energy Electron Diffraction)

A low energy electron beam can be used to obtain a diffraction pattern similar to what is observed with X-ray diffraction. A monoenergetic, low energy electron beam, when diffracted off an ordered surface (see Figure 5.8), will give rise to diffraction spots which can be directly linked to the reciprocal lattice. This image of the reciprocal lattice, in turn, gives information about the real space unit cell. However, it does not provide information about the atomic positions within the unit cell (i.e., basis atom related positions cannot be determined without further analysis). As discussed in elementary solid state texts [19], the direct (Bravais) lattice, which is defined as the (two dimensional) set of vectors **R** for a surface,

$$\mathcal{D_L} = \{\mathbf{R} = n_1\mathbf{a}_1 + n_2\mathbf{a}_2/n_1, \ n_2 \text{ are integers}\} \tag{5.2}$$

where $\mathbf{a}_1$ and $\mathbf{a}_2$ are a chosen set of primitive lattice vectors (which have to be linearly independent). The reciprocal lattice turns out to be somewhat similar, but is spanned by the vectors, $\mathbf{b}_1$ and $\mathbf{b}_2$, given by,

$$\mathcal{R_L} = \{\mathbf{G} = n_1\mathbf{b}_1 + n_2\mathbf{b}_2/n_1, \ n_2 \text{ are integers}\}. \tag{5.3}$$

The primitive vectors of the reciprocal lattice are related to those of the direct lattice through the relation

$$\mathbf{a}_i \cdot \mathbf{b}_j = 2\pi \delta_{ij}. \tag{5.4}$$

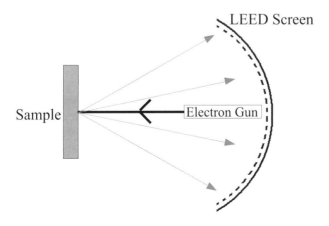

Figure 5.8. LEED setup.

The scattered LEED intensity can be written in terms of a lattice factor $L(\mathbf{G})$ and a structure factor. These contain information about the reciprocal lattice vectors defined above as well as the multiple scattering and arrangement of the atoms inside the unit cell (i.e., the basis atoms at $\mathbf{u}_i \in \mathcal{B}$). The latter introduces another modulation factor to the scattered intensity. These can be expressed as,

$$L(\mathbf{G}) = \left| \sum_{\mathbf{R} \in \mathcal{D}_{\mathcal{L}}} \exp(i\mathbf{G} \cdot \mathbf{R}) \right|^2 \tag{5.5}$$

and

$$S(\mathbf{G}) = \left| \sum_{\mathbf{u}_i \in \mathcal{B}} \exp(i\mathbf{G} \cdot \mathbf{u}_i) \right|^2, \tag{5.6}$$

with the intensity given by the product $L(\mathbf{G})S(\mathbf{G})$. If the basis atoms are different, the structure factor gets modulated by a form factor which introduces those differences to the scattered intensity.

In a real experiment, the electron beam is only able to "sample" the periodic arrangement of the atoms within the first few layers of a surface. LEED is most often used to identify the symmetry and in plane lattice vectors of the surface. What is done usually is to utilize a "trial and error" approach, by calculating the intensities for a model structure and improving it until the calculations agree with experiment [20]. A detailed analysis of the I–V characteristics of the diffracted beam can yield the full surface crystal structure but such a detailed analysis has to be carried out by a specialist. Because of normal incidence, LEED is not generally used for monitoring film growth. However, contrary to the frequently published opinion that LEED is quite insensitive to surface defects, Ref. [22] indicates that it is capable of providing such information.

5.9.   RHEED (Reflection High Energy Electron Diffraction)

RHEED is normally used to monitor and determine the quality of a surface during its growth. A high energy electron beam is reflected from a given surface at glancing angles (in contrast to LEED). The resulting diffraction pattern can be used to identify surface topology, such as steps. In Figures 5.9 and 5.10, such a diffraction pattern and a RHEED setup are shown. If the growth is layer by layer, then there will be a repeating (oscillatory) behavior of an intensity pattern which can be used to determine the number of layers grown, hence the thickness of the film (see Figure 5.11 and Ref. [3]). Also, in layer by layer growth, there will be more steps at near half a monolayer growth, when compared to near completion of a full monolayer. RHEED pattern will be able to differentiate between such topological scenarios. This technique can be also useful for determining the termination layer which is of utmost importance in the growth of more complex layered heterostructures. In addition, in wedge samples when used after growth, RHEED can determine the wedge thickness as a function of position along the wedge.

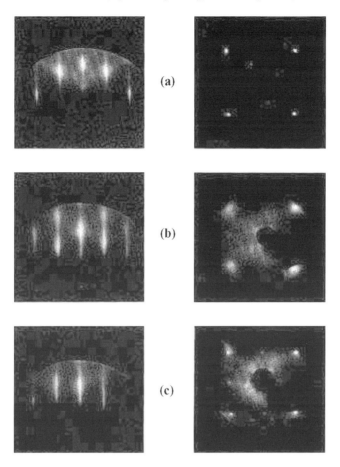

Figure 5.9.   Images from RHEED (left) and LEED (right) patterns for (a) the Ag substrate; (b) 6 ML of Fe on Ag(100); (c) after annealing of the film in (b) at 150 °C. The electron energy of the LEED pictures is $\approx$ 120 eV. This experiment is further discussed in Chapter 4. Reproduced with permission of the American Physical Society from Ref. [21].

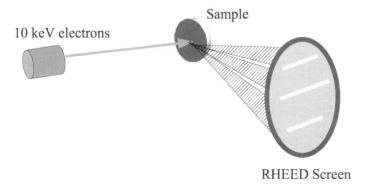

Figure 5.10.   RHEED setup.

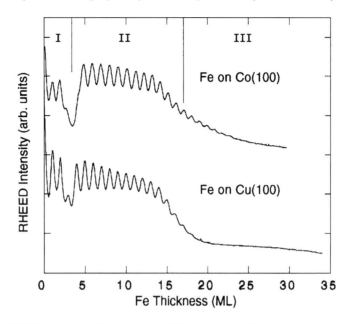

Figure 5.11.   RHEED intensity oscillations in Fe/Co(100) and Fe/Cu(100) during growth of overlayers at room temperature. The similarity of the two sets indicate that the growth modes are quite similar. Reproduced with permission of Elsevier from Ref. [3].

5.10.   ARPES (Angle Resolved Photo-Emission Spectroscopy)

Angle Resolved Photoemission has become an extremely useful, surface sensitive technique to probe electronic states and their dispersions in layered systems. ARPES needs clean and atomically flat sample as well as an ultra high vacuum. With the advent of synchrotron radiation, this technique is even more versatile today. As illustrated in Figure 5.12, a high intensity beam of photons (of known energy and momentum) impinges upon a surface of a given material and ejects electrons from the valence states. When these electrons are detected in the outside vacuum, their kinetic energies can yield information about the original binding energies of the electrons (from conservation of energy). In addition, conservation of parallel momentum, $k_\parallel$ to within a reciprocal lattice vector of the surface reciprocal space, yield information about the **k** vector associated with the electronic state. Ignoring the momentum of the incident photon, we have

$$\mathbf{k}_\parallel^o = \mathbf{k}_\parallel^i + \mathbf{G}_\parallel \tag{5.7}$$

and

$$\left|\mathbf{k}_\parallel^o\right| = \left\{\frac{2m\,E_{kin}}{\hbar^2}\right\}^{1/2} \sin(\theta). \tag{5.8}$$

Here i stands for *inside* and o for *outside* the surface and $\mathbf{G}_\parallel$ is a surface reciprocal lattice vector. For a 2-dimensional ordered lattice, the energy level structure will be a periodic function in the $2-d$ **k**-space, whose repeating unit is the surface Brillouin zone of the 2-d layer. To map out the dispersion E vs. $\mathbf{k}_\parallel$, one measures the kinetic energy (outside) and relate it to the binding energy E_B as,

$$E_{kin} = \hbar\omega - e\phi - E_B. \tag{5.9}$$

By carrying out spin-resolved measurements, it has become possible to obtain spin-polarized band structures along the same lines described here. From the above discussion, it is clear that photoemission can be used to map out a band structure of a given material. Band structure calculations are usually carried out with an effective single particle picture. One could wonder whether there are important many-particle effects that could be detected by a technique such as ARPES. Using many-electron ideas from solid state physics, it is known that the photoemission actually measures the spectral function, $A(\mathbf{k}, \omega)$, which is closely related to the imaginary part of the one particle Green function $G(\mathbf{k}, \omega)$ as,

$$A(\mathbf{k}, \omega) = \frac{1}{\pi}\Im G(\mathbf{k}, \omega). \tag{5.10}$$

The spectral function contains important information related to many-body effects such as the self-energy. Hence ARPES can be utilized to identify many-body induced features that might not be included in an effective one particle theory.

A brief discussion about the photoemission matrix elements are in order here since they affect the photoemission intensity. The transition rate between the initial and the final states of the unperturbed Hamiltonian due to a perturbation $H' \exp(-i\omega t)$ is expressed as

$$R_{i \to j} = (2\pi/\hbar)\left|\langle f|H'|i\rangle\right|^2 \delta(E_f - E_i - \hbar\omega), \tag{5.11}$$

and the perturbation is, of course, due to the electromagnetic field of the incoming radiation. With a proper choice of the gauge, the perturbation term can be simplified as

$$H' = \frac{e}{mc}\mathbf{A} \cdot \mathbf{p} + \frac{e^2}{2mc^2}|\mathbf{A}|^2, \tag{5.12}$$

where e, m and c are fundamental constants representing the electron charge, its mass and the speed of light in a vacuum with $\mathbf{A}$ being the vector potential. The ratio of these constants multiplying the $|\mathbf{A}|^2$ and $\mathbf{A} \cdot \mathbf{p}$ terms is $e/2c$ which makes the second term much smaller and hence the $|\mathbf{A}|^2$ term is usually neglected. Hence the transition rate most commonly used is,

$$R_{i \to j} = \frac{2\pi e^2}{\hbar m^2 c^2}\left|\langle \Psi_f|\mathbf{A} \cdot \mathbf{p}|\Psi_i\rangle\right|^2 \delta(E_f - E_i - \hbar\omega). \tag{5.13}$$

There is also a well known approximation referred to as the "sudden approximation" which must hold for a reasonable interpretation of the photoemission spectra.

Nowadays, ARPES is a well established technique for examining electronic structure of complex systems and there are numerous review articles where detailed descriptions and examples are given (see, for example, Ref. [4]). As mentioned earlier, because of the $\mathbf{k}_\parallel$ construction, this technique is particularly powerful for studies of electronic structure of layered solids, and hence extensively used for thin films and multilayers discussed in this volume. The present generation ARPES has an energy resolution of about 2–10 meV and momentum resolution of about 0.2 degrees. The level of sophistication of such studies is well demonstrated in the figure shown here (Figure 5.13) where both EDCs (Energy Distribution Curves) and MDCs (Momentum Distribution Curves) are shown for an optimally doped cuprate.

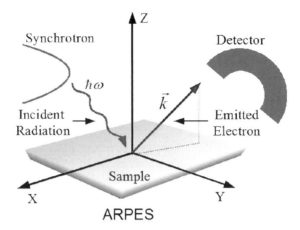

Figure 5.12.   ARPES setup.

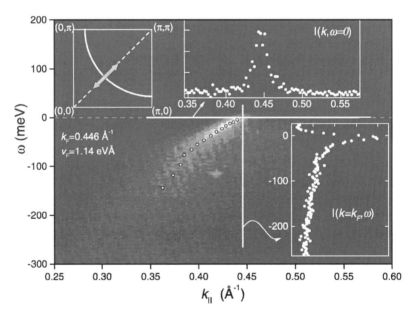

Figure 5.13.   A high resolution ARPES image showing a 2-dimensional spectral plot of emitted intensity from the optimally doped cuprate $Bi_2Sr_2CaCu_2O_{8+\delta}$ along the (π, π) direction of the Brillouin Zone (BZ) as a function of ω the band energy and $\mathbf{k}_{\parallel}$. The photon energy is 21.2 eV and the sample temperature is at 48 K. The insets, from top left, show the BZ cross section, intensity at constant energy ($\omega = 0$ a MDC), and intensity at constant momentum ($\mathbf{k} = \mathbf{k}_F$). Reproduced with the permission of AAAS from Ref. [5].

5.11.   XAS (X-ray absorption spectroscopy)

Energy dependent absorption of X-rays in certain materials will enforce electronic transitions from deep core levels to unoccupied valence states. By monitoring these absorption processes, it is possible to obtain information about the empty valence states as well as information about the neighboring atoms (i.e., local structure). The absorption co-efficient, $\mu(E)$ as a function of the photon energy E can be measured for thin films (absorption length being about 20–100 nm for $3d$ transition metals). Various features in the absorption spectra (such as the K or L edges) contain identifiable fine structure which can be utilized in understanding specific electronic and magnetic properties of materials [26].

The oscillatory behavior of the X-ray absorption coefficient carries information about the local environment of a given atom in the solid state. EXAFS describes the fine structure in the absorption about 30 eV above the X-ray edge, while XANES probes the fine structure in the first 30 eV from the edge position.

5.12. Magnetic Dichroism in XAS

In the above discussion of ARPES, an important underlying assumption is that the electrons can be treated non-relativistically. This is quite valid when dealing with light atoms as well as valence electrons that are well shielded from the nuclei. However, in heavy atoms such as the rare-earths and actinides, relativistic effects become significant. Even in transition metal atoms, when core levels are probed, relativistic effects can play an important role. This is exactly the case when core levels are probed by X-rays. When spin–orbit coupling is strong, the non-relativistic quantum numbers l and s are no longer conserved, i.e., spatial and spin degrees of freedom can no longer be treated independently, and one has to resort to quantum numbers $j = l \pm 1/2$ for a single electron. For example, a core level should not be labeled by its p (i.e., $l = 1$) character, but by quantum numbers $j = 3/2$ or $j = 1/2$. These levels do possess a degeneracy which can be broken by a magnetic field. This leads to further classification of these levels (split) by the magnetic quantum numbers j_z. In the above case, there can be a difference (asymmetry) in dipole transitions from a core level to a continuum free electron state, when negative j_z initial states are compared with positive j_z states. For example, transitions are allowed from $j_z = -3/2, -1/2$ of the core level with right-circularly polarized light while left-circularly polarized light can result in transitions from $j_z = 3/2, 1/2$. The above identification of transitions using circularly polarized light has led to the development of a powerful experimental technique named X-ray Magnetic Circular Dichroism (XMCD), which is capable of determining magnetic properties of complex materials. It also provides element selectivity, due to the local nature of the probe. There are various sum rules that have been derived [23,24], which can be used in the above determinations of magnetic properties. Even when the initial states are valence (band) states, magnetic dichroism can be detected (see, for example, Ref. [25]).

5.13. X-PEEM (X-ray Photoelectron Emission Microscopy)

X-PEEM is a spatially resolved analog of XAS and/or X-ray Dichroism. It makes use of high intensity, tunable X-rays, available from the modern synchrotron facilities, and the interactions of these X-rays with matter [27]. Secondary electrons generated by polarized X-rays incident on a sample (due to the core hole created as in XPS) are imaged by the lenses to form a magnified picture of the electron yield distribution. (The secondary

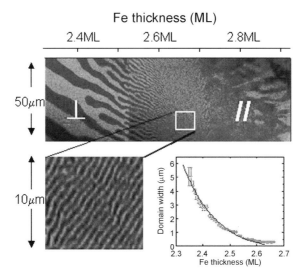

Figure 5.14.   A sample of PEEM results: (a) Image of the magnetic domains of Fe/Ni(5 ML)/Cu(001). The stripe domain width decreases as the iron thickness increases to the spin reorientation transition point at $d_{Fe} \simeq 2.7$ ML. (b) A zoom-in image of the magnetic stripes in the box of (a). (c) Stripe domain width vs. Fe thickness. Reproduced with kind permission of Y.Z. Wu. Also published in Ref. [28].

electron yield is clearly proportional to X-ray absorption.) There are local variations in electron emission (due to local variations in X-ray absorption) that will lead to an image with topographical contrast. This contrast can arise from several factors that might be associated with chemical, electronic, magnetic and lateral inhomogeneities depending on the photoabsorption process. The image contains sufficient information with regard to the electronic and magnetic states of the sample surface and near surface region and comparisons of different images give useful information. This is a technique that is surface sensitive (similar to surface sensitivity of XAS) since the secondary electrons, responsible for creating the image, have a mean free path of a few nanometers, although the X-rays have a much larger probing depth. The mean free path depends strongly on the material but is in the range of 3–10 nm. Note that in PEEM, a large number of electrons are sampled (to form the image) in contrast to standard photoemission experiments (UPS or ARPES), where electrons are sent through a velocity selector. (See Figures 5.14 and 5.15 for sample PEEM images from Cu/Co/Cu multilayers.)

5.14.   SPLEEM (Spin Polarized Low Energy Electron Microscopy)

SPLEEM is a low voltage electron microscopy for the study of surfaces and interfaces. The electron beam is spin polarized and sensitive to magnetic domains and orientations. It has been traditionally used for magnetic imaging with spatial contrast. Recently, it has been shown that the reflected intensity is related to the unoccupied part of the density of states, offering a new way to probe electronic structure at a lateral scale of ~ 50 nm. Due to low energy electrons used, SPLEEM is highly surface sensitive.

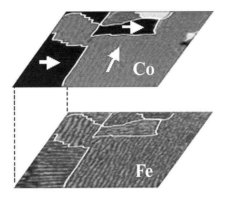

Figure 5.15.   Co and Fe domain images of Co/Cu(6.6 ML)/Fe(2.7 ML)/Ni(5 ML)/Cu(001). Fe/Ni stripe direction is aligned with the in-plane Co magnetization (shown by arrows) through the magnetic interlayer coupling. Reproduced with kind permission of Y. Z. Wu. Also published in Ref. [28].

Figure 5.16 shows a SPLEEM setup where the incident, spin-polarized electron is directed at the sample at normal incidence. The reflected electron moves through the lens and forms an image similar to what is shown in this figure. The electron energy can be tuned by the sample voltage and it is possible to alternate the electron spins and measure spin-dependent reflectivities as shown in Figure 5.16. In Ref. [29], a SPLEEM study of Cu/Co multilayers has been reported where images of the Cu overlayers show magnetic domains of the Co underlayer with the domain contrast oscillating as a function of electron energy as well as the Cu overlayer thickness (not shown here). These oscillations have been attributed to the spin-dependent electron reflectivity from the Cu/Co interface resulting in Fabry–Pérot type electron interference [29].

5.15.   Andreev reflection

We have discussed reflection and transmission of electrons in metallic multilayer systems from a simple-minded QW viewpoint. What happens when one of the components of such a system is a superconductor? This in fact turns out to be a question that Landau had asked one of his students, Andreev, to look into. This remarkable phenomenon associated with such reflections from a superconductor are now referred to as Andreev reflection. On the metallic side, there are many quasi-particle states at the Fermi level while on the superconducting side, there is a gap in the density of states which stabilizes the Cooper pairs. Experimentally, it is known that a current can pass through such a junction (consisting of a normal metal (NM) and a superconductor (SC)) at subgap voltages. When this problem is analyzed in some detail, it turns out that the electrons (or quasi-particles) in the NM with subgap energy cannot enter the SC region directly. What is thought to happen is that an electron (say with subgap energy) incident from the NM side forms a Cooper pair with an electron below the Fermi level in the NM and leaves for the condensate (i.e., SC). The latter electron leaves a hole behind in the NM side which is called the "Andreev-reflected hole". In the reverse process, a Cooper pair can pounce on a hole in the vicinity of the NM/SC interface with one electron (from the Cooper pair) filling the hole and the other moving into the NM region.

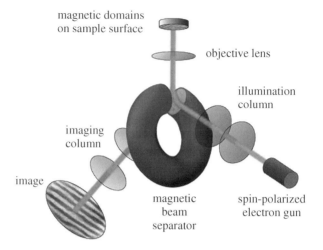

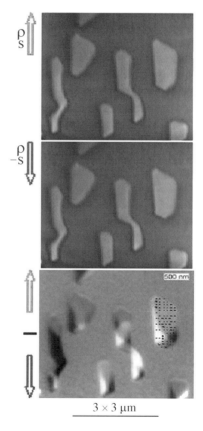

Figure 5.16.   SPLEEM experimental setup and spin-polarized images. Reproduced with kind permission from A.K. Schmid (private communication).

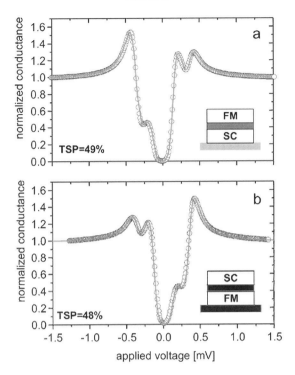

Figure 5.17.   Conductance curves (open symbols) and corresponding fits (lines) for $Co_{84}Fe_{16}$ in (a) normal and (b) inverted structure in an applied field of $\approx$ 1 Tesla. In the normal structure, the SC forms the bottom electrode while in the inverted structure, the ferromagnet (FM) is on the bottom. Tunneling Spin Polarization (TSP) extracted from fits to the data are included and are the same for both structures within the experimental uncertainty. Reproduced from [30] with permission from © 2003 IEEE.

Andreev reflection results in an increased conductance for bias voltages within the superconducting gap [31]. When the metal is ferromagnetic (see Figure 5.17), the current becomes spin-polarized. The current is suppressed to a degree that reflects the spin polarization. Such a device, referred to as a magnetic tunnel junction, can be used to explore the tunneling spin polarization in layered systems.

References

[1] E.R. Moog, C. Liu, S.D. Bader, J. Zak, Phys. Rev. B **39**, 6949 (1989).

[2] S.D. Bader, J. Magn. Magn. Mater. **100**, 440 (1991).

[3] Z.Q. Qiu, S.D. Bader, J. Magn. Magn. Mater. **200**, 664 (1999).

[4] A. Damascelli, Physica Scripta T **109**, 61 (2004).

[5] T. Valla, A.V. Fedorov, P.D. Johnson, B.O. Wells, S.L. Hulbert, Q. Li, G.D. Gu, N. Koshizuka, Science **285**, 2110 (1999).

[6] M.P. Seah, C.P. Hunt, Rev. Sci. Instrum. **59**, 217 (1988).

[7] H. Landskron, G. Schmidt, K. Heinz, K. Müller, C. Stuhlmann, U. Beckers, M. Wuttig, H. Ibach, Surf. Sci. **256**, 115 (1991).

[8] C.L. Briant, R.P. Messmer (Eds.), Auger Electron Spectroscopy, Treatise on Materials Science and Technology, vol. 30, Academic Press (1988).

[9] C. Argile, G.E. Rhead, Surface Science Reports **10**, 277 (1989).

[10] D. Sander, W. Pan, S. Oauzi, J. Kirschner, W. Meyer, M. Krause, S. Mueller, L. Hammer, K. Heinz, Phys. Rev. Lett. **93**, 247203 (2004).

[11] J.H. Griffiths, Nature **168**, 670 (1946).

[12] B. Heinrich, J.F. Cochran, Adv. Phys. **42**, 523 (1993).

[13] B. Hillebrands, G. Güntherodt, in: B. Heinrich, J.A.C. Bland (Eds.), Ultrathin Magnetic Structures II, Springer (1994).

[14] G. Binnig, H. Rohrer, Ch. Gerber, E. Weibel, Phys. Rev. Lett. **49**, 57 (1982).

[15] F.J. Giessibl, Rev. Mod. Phys. **75**, 949 (2003).

[16] G.L. Squires, Introduction to the Theory of Thermal Neutron Scattering (Dover, New York, 1997).

[17] H. Dachs (Ed.), Neutron Diffraction, Springer-Verlag, New York (1978).

[18] U. Gonser (Ed.), Mössbauer Spectroscopy, Springer-Verlag, New York, Heidelberg, Berlin (1975).

[19] N. Ashcroft, D. Mermin, Solid State Physics (Saunders College, 1976).

[20] M. Wuttig, X. Liu, Ultrathin Metal Films (Springer-Verlag, Berlin, Heidelberg, 2004).

[21] Z.Q. Qiu, J. Pearson, S.D. Bader, Phys. Rev. Lett. **70**, 1006 (1993).

[22] M. Henzler, See for example, in: H. Ibach (Ed.), Electron Spectroscopy for Surface Analysis, Springer-Verlag (1977).

[23] B.T. Thole, et al., Phys. Rev. Lett. **68**, 1943 (1992).

[24] P. Carra, et al., Phys. Rev. Lett. **70**, 694 (1993).

[25] W. Kuch, M. Schneider, Rep. Progress in Phys. **64**, 147 (2001).

[26] J.J. Rehr, R.C. Albers, Rev. Mod. Phys. **72**, 621 (2000).

[27] A. Scholl, H. Ohldag, F. Nolting, J. Stohr, H.A. Padmore, Rev. Sci. Intrum. **73**, 1362 (2002).

[28] Y.Z. Wu, et al., Phys. Rev. Lett. **93**, 117205 (2004).

[29] Y.Z. Wu, A.K. Schmid, M.S. Altman, X.F. Jin, Z.Q. Qiu, Phys. Rev. Lett. **94**, 027201 (2005).

[30] S.S.P. Parkin, X. Jian, C. Kaiser, A. Panchula, K. Roche, M. Samant, Proceedings of the IEEE **91**, 661 (2003).

[31] G.E. Blonder, M. Tinkham, T.M. Klapwijk, Phys. Rev. B **25**, 4515 (1982).

Chapter 6

Generalized Kohn–Sham Density Functional Theory via Effective Action Formalism

6.1. Introduction

Modern density functional theory is one of the most popular and versatile methods available to condensed matter physicists as well as computational chemists to study ground state properties of various condensed matter systems. First principles studies of metallic multilayers and other systems discussed in some of the previous chapters are mostly based on various approximate density or spin-density functionals. Here we examine the foundation of density functional theory from a rigorous point of view. Density functional theory [1] allows one to study the ground state properties of the many-body system in terms of the expectation value of the particle-density operator. In principle, it offers the possibility of finding the ground state energy E_g by minimizing the energy functional that depends on the density only:

$$E_g = \min_n E[n]. \tag{6.1}$$

Similar theories can be formulated in terms of the expectation values of a spin-density operator or current-density operator, known as spin-density [2] and current-density functional theory [3]. In general, one can imagine a description of a many-body system in terms of the expectation value of any other suitable operator. Such a general description can be elegantly presented via the effective action formalism [4,5], thus leading to a *generalized* density-functional theory – a theory that allows a description of many-body systems in terms of the expectation value of any suitable operator.

Proving the existence of such a theory is not a trivial matter. In fact, the conventional density-functional theory relies heavily on the theorem of Hohenberg and Kohn [1], which shows that there exists a *unique* description of a many-body system in terms of the expectation value of the particle-density operator. Finite-temperature extension of this theorem was given by Mermin [8]. Similar results have been obtained for spin-density and current-density-functional theory [2,3]. Clearly, it is vital to establish the corresponding existence theorems when constructing a generalized density-functional theory. In the framework of the effective action formalism, the proof of existence was given only in a diagrammatic sense and was tightly tailored to the particular features of the particle-density based description of a nonrelativistic many-electron system [4]. In this work we present a different resolution of this important issue. Our formulation is valid for a general case and does not rely on any perturbative expansions. Not only does this offer an alternative proof of the previous results for density-, spin-density-, current-density-functional theories, but it also provides a rigorous foundation for a generalized density-functional theory (see, for example, recent related work in Ref. [6,7]).

However important the formal question of existence is, it is of little help for the actual construction of the required density or other functionals. For example, to make any practical use of the conventional density-functional theory, an explicit (perhaps approximate) expression for the energy functional (6.1) is necessary. An important contribution here was made by Kohn and Sham [9]. They proposed a certain decomposition of the energy functional, which for a typical nonrelativistic many-fermion system,

$$
\hat{H} = \int d\mathbf{x}\, \hat{\psi}^\dagger(\mathbf{x}) \left(-\frac{1}{2m}\nabla^2 + \upsilon_{ion}(\mathbf{x}) \right) \hat{\psi}(\mathbf{x})
$$
$$
+ \frac{e^2}{2} \iint \frac{\hat{\psi}^\dagger(\mathbf{x})\hat{\psi}^\dagger(\mathbf{y})\hat{\psi}(\mathbf{y})\hat{\psi}(\mathbf{x})}{|\mathbf{x}-\mathbf{y}|}\, d\mathbf{x}\, d\mathbf{y}, \tag{6.2}
$$

takes the form

$$
E[n] = T_s[n] + \int \upsilon_{ion}(\mathbf{x})n(\mathbf{x})\, d\mathbf{x}
$$
$$
+ \frac{e^2}{2} \iint \frac{n(\mathbf{x})n(\mathbf{y})}{|\mathbf{x}-\mathbf{y}|}\, d\mathbf{x}\, d\mathbf{y} + E_{xc}[n]. \tag{6.3}
$$

Here $T_s[n]$ is the kinetic energy of an auxiliary system of noninteracting fermions that yields the ground state density $n(\mathbf{x})$, and $E_{xc}[n]$ is the exchange-correlation energy. Once the approximation for $E_{xc}[n]$ has been decided, the minimization of the functional (6.3) leads to the familiar Kohn–Sham single-particle equations [10]. This approach represents the so-called Kohn–Sham density-functional theory.

A natural question now arises: is there an analog of Kohn–Sham density-functional theory in the effective action formalism? In the original work of Fukuda, Kotani, Suzuki and Yokojima [4], the relationship between the two methods was not established. In fact regarding a Kohn–Sham single particle equation, Ref. [4] states that "such an equation can be written down but its physical meaning is not clear". Understanding the place (i.e., the physical meaning) of Kohn–Sham theory in the effective action formalism is of paramount importance for a number of reasons. A vast majority of first-principles calculations are based on the Kohn–Sham method [11,12]. The remarkable success of these calculations points to the fact that this seemingly ad hoc decomposition provides a very good approximation of the energy functional. As we show in this chapter, the connection between the effective action formalism and the Kohn–Sham method can be rigorously established via the inversion method [13]. The realization of this fact immediately leads to a generalized Kohn–Sham theory.

A distinct feature of the effective action formulation of generalized Kohn–Sham theory is that it provides a systematic way of calculating the required functionals. For example, in the case of a particle-density based description of a nonrelativistic many-electron system, it leads to a set of simple diagrammatic rules for constructing the exchange-correlation functional entirely in terms of Kohn–Sham derived quantities. We construct the first few orders of the exchange-correlation functional, comment on the local density approximation, and discuss our results as compared to other methods [11,14,15].

Applications of the presented formalism is not restricted to the exchange-correlation functional only. Earlier, it was demonstrated that this method is capable of establishing rigorous Kohn–Sham density-functional formulation of one-electron propagators [16]. We briefly discuss the main results of that work and compare it with the existing strategies for construction of one-electron propagators [17–19]. Lastly, we analyze the excitation energies within the effective action formalism and comment on the relationship with similar results [20,21] obtained via time-dependent density-functional theory [22].

6.2. Effective action functional

The effective action formalism in the context of density-functional theory was discussed by Fukuda, Kotani, Suzuki and Yokojima [4]. Here we describe the main features of this method and prove the generalized existence theorems. We start by defining the functional $W[J]$ as

$$e^{-\beta W[J]} = Tr\left(e^{-\beta(\hat{H}+J(1)\hat{Q}(1))}\right). \tag{6.4}$$

Here $\hat{H}$ denotes the Hamiltonian of the system under consideration. Parameter β can be identified with inverse temperature. $\hat{Q}$ is the operator whose expectation value will serve as a main variable of the theory, and J is the external source coupled to it. Both $\hat{H}$ and $\hat{Q}$ are assumed to be time-independent. Summation over repeated indices is assumed, and the notation $J(1)\hat{Q}(1)$ embodies all the necessary summations and integrations. For example, to formulate the theory in terms of the expectation value of the particle-density operator $\hat{n}(\mathbf{x})$, we choose $\hat{Q} = \hat{n}(\mathbf{x})$ and

$$J(1)\hat{Q}(1) \equiv \int d\mathbf{x}\, J(\mathbf{x})\hat{n}(\mathbf{x}). \tag{6.5}$$

Thermal expectation value of $\hat{Q}$,

$$Q(1) = \frac{Tr(\hat{Q}(1)e^{-\beta(\hat{H}+J(1')\hat{Q}(1'))})}{Tr(e^{-\beta(\hat{H}+J(1')\hat{Q}(1'))})}, \tag{6.6}$$

can be written in terms of $W[J]$ as,

$$Q(1) = \frac{\delta W[J]}{\delta J(1)}. \tag{6.7}$$

Denoting the set of allowable external sources as $\mathcal{J}$ and the set of all generated expectation values of Q as $\mathcal{Q}$, we can establish a map $\mathcal{J} \to \mathcal{Q}$. We assume that for a given element of $\mathcal{J}$ there corresponds only one element of $\mathcal{Q}$, in other words, given the external source we can unambiguously establish the expectation value that it generates. Whether the converse of this statement is true remains to be proven. The following property of the functional $W[J]$ is of fundamental importance.

THEOREM 1. *The functional $W[J]$ is strictly concave, i.e., for any α, $0 < \alpha < 1$, and $J \neq J'$*

$$W\left[\alpha J + (1-\alpha)J'\right] > \alpha W[J] + (1-\alpha)W[J']. \tag{6.8}$$

Based on strict concavity of $W[J]$ one can prove the following result.

COROLLARY 1. *The map* $\mathcal{J} \to \mathcal{Q}$ *is one-to-one.*

The proofs of these statements are given at the end of this chapter. Corollary 1 guarantees that the functional relationship between J and Q can be inverted:

$$\frac{\delta W[J]}{\delta J(1)} = Q(1) \Rightarrow J = J[Q]. \tag{6.9}$$

When $\hat{Q} = \hat{n}(\mathbf{x})$, Corollary 1 represents an alternative proof to the theorem of Mermin [8].

The functional $W[J]$ provides a description of the physical system in terms of the external probe J. We, on the other hand, want the description in terms of Q. The change of variables from J to Q can be accomplished via a functional Legendre transformation [23]. This leads to the definition of the effective action functional:

$$\Gamma[Q] = W[J] - J(1')Q(1'). \tag{6.10}$$

Here, J is assumed to be a functional of Q from Equation (6.9). The functional $\Gamma[Q]$ possesses the following important property.

PROPOSITION 1. *The effective action functional $\Gamma[Q]$ defined on the set $\mathcal{Q}$ is strictly convex.*

Proof of this statement is given at the end of this chapter. Differentiating (6.10) with respect to Q, we obtain

$$\frac{\delta \Gamma[Q]}{\delta Q(1)} = -J(1). \tag{6.11}$$

Since our original system is recovered when $J = 0$, we arrive at the important variational principle: the functional $\Gamma[Q]$ reaches a minimum at the exact expectation value of $\hat{Q}$,

$$\left(\frac{\delta \Gamma[Q]}{\delta Q} \right)_{Q=Q_g} = 0. \tag{6.12}$$

In the zero temperature limit, $\beta \to \infty$, Q_g represents the exact ground state expectation value of $\hat{Q}$ and $\Gamma[Q_g]$ equals the exact ground state energy. Obviously, at finite temperatures

$$\Gamma[Q_g] = -\frac{1}{\beta} \ln Tr\left(e^{-\beta \hat{H}} \right), \tag{6.13}$$

$$Q_g(1) = \frac{Tr(\hat{Q}(1)e^{-\beta \hat{H}})}{Tr(e^{-\beta \hat{H}})}. \tag{6.14}$$

The effective action formalism furnishes a rigorous formulation of generalized density-functional theory. The existence of $\Gamma[Q]$ is guaranteed by Corollary 1. When $\hat{Q}$ is a particle-density operator, we obtain conventional density-functional theory; when $\hat{Q}$ stands for a spin-density operator, we have spin-density-functional theory [2]; when $\hat{Q}$ is a current operator, we obtain current-density-functional theory, [3], etc.

To implement this formally exact method, an approximation of the effective action functional $\Gamma[Q]$ is required. For a particle-density based description of a nonrelativistic many-electron system the effective action functional can be approximated via the auxiliary field method [4]. However, this method does not lead [4] to Kohn–Sham version of density-functional theory. The relationship between the effective action formalism and Kohn–Sham density-functional theory can be established via the inversion method [13].

6.3. Generalized Kohn–Sham theory via the inversion method

Consider the following general Hamiltonian

$$\hat{H} = \hat{H}_0 + \lambda \hat{H}_{int}, \tag{6.15}$$

which depends on the coupling constant λ as a parameter. The same is true for the effective action functional

$$\Gamma = \Gamma[Q, \lambda]. \tag{6.16}$$

Clearly Q, λ are to be considered as two independent variables. Note, however, that this does not prevent the exact expectation value Q_g from depending on λ: this dependence is fixed by the variational principle

$$\left(\frac{\delta \Gamma[Q, \lambda]}{\delta Q} \right)_{Q_g} = 0. \tag{6.17}$$

The functional $\Gamma[Q, \lambda]$ is defined as,

$$\Gamma[Q, \lambda] = W[J, \lambda] - J(1')Q(1'), \tag{6.18}$$

where J is functional of Q and λ. This functional dependence is provided by the equation

$$\frac{\delta W[J, \lambda]}{\delta J(1)} = Q(1). \tag{6.19}$$

The inversion method [5,13] proceeds by expanding all the quantities in Equation (6.18) in terms of λ;

$$
\begin{aligned}
J[Q, \lambda] &= J_0[Q] + \lambda J_1[Q] + \lambda^2 J_2[Q] + \cdots, \\
W[J, \lambda] &= W_0[J] + \lambda W_1[J] + \lambda^2 W_2[J] + \ldots, \\
\Gamma[Q, \lambda] &= \Gamma_0[Q] + \lambda \Gamma_1[Q] + \lambda^2 \Gamma_2[Q] + \cdots.
\end{aligned}
\tag{6.20}
$$

Comparison of the two sides in Equation (6.18) for different orders of λ,

$$\sum \lambda^i \Gamma_i[Q] = \sum \lambda^i W_i \left[\sum \lambda^k J_k[Q] \right] - \sum \lambda^i J_i(1)Q(1), \tag{6.21}$$

leads to the expression for $\Gamma_l[Q]$,

$$
\Gamma_l[Q] = W_l[J_0] + \sum_{k=1}^{l} \frac{\delta W_{l-k}[J_0]}{\delta J_0(1)} J_k(1) - J_l(1)Q(1)
$$

$$
+ \sum_{m=2}^{l} \frac{1}{m!} \sum_{\substack{k_1+\cdots+k_m \leqslant l \\ k_1,\ldots,k_m \geqslant 1}} \frac{\delta^m W_{l-(k_1+\cdots+k_m)}[J_0]}{\delta J_0(1) \cdots \delta J_0(m)} J_{k_1}(1) \cdots J_{k_m}(m).
$$

Functionals $\{W_l[J_0]\}$ and its derivatives are assumed to be known; they can usually be obtained via standard many-body perturbation techniques (specific examples will be given in the next section). Since Q and λ are considered to be independent, it follows from Equation (6.11) that functionals $\{J_k[Q]\}$ can be obtained using,

$$\frac{\delta \Gamma_k[Q]}{\delta Q(1)} = -J_k(1).\tag{6.22}$$

Consider the zeroth order term,

$$\Gamma_0[Q] = W_0[J_0] - J_0(1)Q(1).\tag{6.23}$$

Using Equation (6.22)

$$-J_0(1) = \frac{\delta W_0[J_0]}{\delta J_0(1')}\frac{\delta J_0(1')}{\delta Q(1)} - J_0(1) - Q(1')\frac{\delta J_0(1')}{\delta Q(1)}$$

$$\Rightarrow \left(\frac{\delta W_0[J_0]}{\delta J_0(1')} - Q(1')\right)\frac{\delta J_0(1')}{\delta Q(1)} = 0.\tag{6.24}$$

Strict convexity of $\Gamma_0[Q]$ (see Proposition 1) prohibits $(\delta J_0(1')/\delta Q(1))$ from having zero eigenvalues. Thus we obtain that J_0 obeys the equation:

$$Q(1) = \frac{\delta W_0[J_0]}{\delta J_0(1)}.\tag{6.25}$$

Hence J_0 is determined as a potential which generates the expectation value Q in the *noninteracting* ($\lambda = 0$) system. Notice that the same exact notion appears in Kohn–Sham formalism [10]. We refer to this noninteracting system as Kohn–Sham (KS) system and J_0 as Kohn–Sham potential. Equation (6.25) allows one to simplify the expression for $\Gamma_l[Q]$, which now becomes,

$$\Gamma_l[Q] = W_l[J_0] - \delta_{l,0}J_0(1)Q(1) + \sum_{k=1}^{l-1}\frac{\delta W_{l-k}[J_0]}{\delta J_0(1)}J_k(1)$$

$$+ \sum_{m=2}^{l}\frac{1}{m!}\sum_{\substack{k_1,\ldots,k_m \geqslant 1}}^{k_1+\cdots+k_m \leqslant l}\frac{\delta^m W_{l-(k_1+\cdots+k_m)}[J_0]}{\delta J_0(1)\cdots\delta J_0(m)}J_{k_1}(1)\cdots J_{k_m}(m).\tag{6.26}$$

The important message here is that the expression for $\Gamma_l[Q]$ involves only lower order functionals ($l-1, l-2, \ldots, 0$). Thus starting with J_0 we can determine $\Gamma_1[Q]$ as

$$\Gamma_1[Q] = W_1[J_0].\tag{6.27}$$

From $\Gamma_1[Q]$ we can find J_1 as,

$$J_1(1) = -\frac{\delta \Gamma_1[Q]}{\delta Q(1)} = -\frac{\delta \Gamma_1[Q]}{\delta J_0(1')}\frac{\delta J_0(1')}{\delta Q(1)},$$

which can be written in the alternate form

$$J_1(1) = \mathcal{D}(1,2)\frac{\delta W_1[J_0]}{\delta J_0(2)},\tag{6.28}$$

where the inverse propagator is defined as,

$$\mathcal{D}(1,2) = -\frac{\delta J_0(2)}{\delta Q(1)} = -\left(\frac{\delta^2 W_0[J_0]}{\delta J_0(1)\delta J_0(2)}\right)^{-1}. \tag{6.29}$$

Once J_1 is known, we can find $\Gamma_2[Q]$:

$$\Gamma_2[Q] = W_2[J_0] + \frac{\delta W_1[J_0]}{\delta J_0(1)} J_1(1) + \frac{1}{2}\frac{\delta^2 W_0[J_0]}{\delta J_0(1)\delta J_0(2)} J_1(1)J_1(2), \tag{6.30}$$

or using Equation (6.28)

$$\Gamma_2[Q] = W_2[J_0] + \frac{1}{2}\frac{\delta W_1[J_0]}{\delta J_0(1)}\mathcal{D}(1,2)\frac{\delta W_1[J_0]}{\delta J_0(2)}. \tag{6.31}$$

From $\Gamma_2[Q]$ the expression for J_2 follows as,

$$J_2(1) = \mathcal{D}(1,2)\frac{\delta W_2[J_0]}{\delta J_0(2)} + \mathcal{D}(1,2)\frac{\delta^2 W_1[J_0]}{\delta J_0(2)\delta J_0(2')}J_1(2')$$
$$+ \frac{1}{2}\mathcal{D}(1,2)\frac{\delta^3 W_0[J_0]}{\delta J_0(2)\delta J_0(3)\delta J_0(4)}J_1(3)J_1(4). \tag{6.32}$$

This, in turn, leads to $\Gamma_3[Q]$ and so on. In this hierarchical fashion one can consistently determine $\Gamma[Q,\lambda]$ to any required order;

$$\Gamma_0 \to J_0 \to \Gamma_1 \to J_1 \to \Gamma_2 \to J_2 \to \cdots. \tag{6.33}$$

The important point here is that all higher orders are completely determined by the Kohn–Sham potential J_0. Let us now apply the variational principle to our expansion of $\Gamma[Q,\lambda]$:

$$\frac{\delta\Gamma[Q,\lambda]}{\delta Q(1)} = 0. \tag{6.34}$$

Since

$$\frac{\delta\Gamma_0[Q]}{\delta Q(1)} = -J_0(1), \tag{6.35}$$

we have the following important result

$$J_0(1) = \frac{\delta\Gamma_{int}[Q]}{\delta Q(1)} = -\mathcal{D}(1,1')\frac{\delta\Gamma_{int}[Q]}{\delta J_0(1')}, \tag{6.36}$$

where

$$\Gamma_{int}[Q] = \sum_{i=1}\lambda^i \Gamma_i[Q]. \tag{6.37}$$

Since at any order the effective action functional is completely determined by Kohn–Sham potential J_0, we arrive at generalized Kohn–Sham self-consistent method:

1. Choose the approximation for $\Gamma_{int}[Q]$ (one obvious choice is to truncate the expansion at some order).
2. Start with some reasonable guess for the Kohn–Sham potential J_0.
3. Calculate $\Gamma_{int}[Q]$.

4. Determine new Kohn–Sham potential J_0 via Equation (6.36).
5. Repeat from step 3 until self-consistency is achieved.

   The formalism described above can be applied to any general case and provides a rigorous basis for generalized Kohn–Sham theory. Practical implementation of the self-consistent procedure obviously depends on the particular Hamiltonian under consideration and the choice of the operator $\hat{Q}$. In the next section we apply this method to the case of a particle-density based description of nonrelativistic many-electron system.

6.4.   Kohn–Sham density-functional theory

6.4.1.   Derivation of Kohn–Sham decomposition

Let us consider a typical nonrelativistic many-electron system described by the Hamiltonian (6.2) and develop the description in terms of the particle-density operator:

$$\hat{Q}(1) = \hat{\psi}^{\dagger}(\mathbf{x})\hat{\psi}(\mathbf{x}) \equiv \hat{n}(\mathbf{x}). \tag{6.38}$$

For convenience, the spin degrees of freedom are suppressed here (those can be easily recovered if necessary). The role of the coupling constant λ is played by e^2. We now evaluate the effective action functional $\Gamma[n]$ using the inversion method [13] described in the previous section. Coupling constant expansion of the functional $W[J_0, e^2]$,

$$W[J_0, e^2] = W_0[J_0] + e^2 W_1[J_0] + e^4 W_2[J_0] + \cdots, \tag{6.39}$$

can be conveniently generated using path integral representation [25]

$$e^{-\beta W[J]} = \int D\psi^{\dagger} D\psi \, e^{-S[\psi^{\dagger}, \psi] - \int J(\mathbf{x})\psi^{\dagger}(x)\psi(x)\,dx}. \tag{6.40}$$

Here

$$S[\psi^{\dagger}, \psi] = \int dx \, \psi^{\dagger}(x) \left[\frac{\partial}{\partial \tau} - \frac{\nabla^2}{2m} + \upsilon_{ion}(\mathbf{x}) \right] \psi(x)$$
$$+ \frac{e^2}{2} \iint dx \, dx' \, \psi^{\dagger}(x)\psi^{\dagger}(x')u(x - x')\psi(x')\psi(x), \tag{6.41}$$

$$u(x - x') = \frac{\delta(\tau - \tau')}{|\mathbf{x} - \mathbf{x}'|}, \tag{6.42}$$

$$\int dx \equiv \int_0^{\beta} d\tau \int d\mathbf{x}, \tag{6.43}$$

and $\psi^{\dagger}$, ψ denote Grassmann fields [25].

   The zeroth order term is given by,

$$W_0[J_0] = -\frac{1}{\beta} \sum_i \ln\left(1 + e^{-\beta\varepsilon_i}\right), \tag{6.44}$$

where ε_i's denote single-particle energies of Kohn–Sham noninteracting system:

$$\left(-\frac{\nabla^2}{2m} + v_{ion}(\mathbf{x}) + J_0(\mathbf{x})\right)\varphi_i(\mathbf{x}) = \varepsilon_i \varphi_i(\mathbf{x}), \tag{6.45}$$

$$n(\mathbf{x}) = \sum_i n_i |\varphi_i(\mathbf{x})|^2, \quad n_i = \left(e^{\beta\varepsilon_i} + 1\right)^{-1}. \tag{6.46}$$

First order term is given by,

$$W_1[J_0] = \frac{1}{2\beta} \bigcirc \cdots \bigcirc - \frac{1}{2\beta} \bigcirc \tag{6.47}$$

Here solid lines denote Matsubara Green's function of Kohn–Sham noninteracting system

$$\mathcal{G}_0(\mathbf{x}\tau, \mathbf{x}'\tau') = \begin{cases} \tau > \tau', & \sum \varphi_i(\mathbf{x})\varphi_i^*(\mathbf{x}')e^{-\varepsilon_i(\tau-\tau')}(n_i - 1), \\ \tau \leqslant \tau', & \sum \varphi_i(\mathbf{x})\varphi_i^*(\mathbf{x}')e^{-\varepsilon_i(\tau-\tau')}n_i, \end{cases} \tag{6.48}$$

and dotted line stands for the Coulomb interaction $u(x - x')$. The explicit expression for $W_1[J_0]$ is given by,

$$W_1[J_0] = \frac{1}{2\beta} \iint \mathcal{G}_0(x, x)u(x - x')\mathcal{G}_0(x', x')\, dx\, dx'$$
$$- \frac{1}{2\beta} \iint \mathcal{G}_0(x, x')u(x - x')\mathcal{G}_0(x', x)\, dx\, dx'. \tag{6.49}$$

The second order term is

$$W_2[J_0] = \frac{1}{4\beta} \bigotimes - \frac{1}{4\beta} \bigcirc\bigcirc$$

$$+ \frac{1}{2\beta} \bigcirc - \frac{1}{\beta} \bigcirc\bigcirc$$

$$+ \frac{1}{2\beta} \bigcirc\cdots\bigcirc\cdots\bigcirc \tag{6.50}$$

Similarly, one can generate higher order terms. Let us now calculate the first few orders of the effective action functional. Using Equation (6.23), the zeroth order correction is given by,

$$\Gamma_0[n] = -\frac{1}{\beta} \sum_i \ln\left(1 + e^{-\beta\varepsilon_i}\right) - \int J_0(\mathbf{x})n(\mathbf{x})\, d\mathbf{x}. \tag{6.51}$$

In the zero temperature limit, $\beta \to \infty$, this transforms to

$$\Gamma_0[n] = \sum_{i=1}^{N} \varepsilon_i - \int J_0(\mathbf{x})n(\mathbf{x})\, d\mathbf{x}, \tag{6.52}$$

or

$$\Gamma_0[n] = T_0[n] + \int v_{ion}(\mathbf{x})n(\mathbf{x})\,d\mathbf{x}. \tag{6.53}$$

Hence at zeroth order, the effective action functional is given by the sum of kinetic energy $T_0[n]$ and ion-potential energy of the *Kohn–Sham noninteracting system*. From Equation (6.27), the first order correction is given by,

$$\Gamma_1[n] = W_1[J_0]. \tag{6.54}$$

The expression for $W_1[J_0]$ has been presented earlier (see Equation (6.47)). In the zero temperature limit

$$\Gamma_1[n] = \frac{1}{2}\iint \frac{n(\mathbf{x})n(\mathbf{x}')}{|\mathbf{x} - \mathbf{x}'|}\,d\mathbf{x}\,d\mathbf{x}' - \frac{1}{2}\iint \frac{n(\mathbf{x}, \mathbf{x}')n(\mathbf{x}', \mathbf{x})}{|\mathbf{x} - \mathbf{x}'|}\,d\mathbf{x}\,d\mathbf{x}', \tag{6.55}$$

where $n(\mathbf{x}, \mathbf{x}') = \sum \varphi_i(\mathbf{x})\varphi_i^*(\mathbf{x}')$. Here, the first term represents the classical Hartree energy and the second term represents the exchange energy. Both are evaluated with respect to Kohn–Sham noninteracting system.

Postponing the evaluation of the second order term until the next section, let us summarize the results we have obtained. It is clear that in the zero temperature limit the expansion for the effective action functional,

$$\Gamma[n] = T_0[n] + \int v_{ion}(\mathbf{x})n(\mathbf{x})\,d\mathbf{x} + \frac{e^2}{2}\iint \frac{n(\mathbf{x})n(\mathbf{x}')}{|\mathbf{x} - \mathbf{x}'|}\,d\mathbf{x}\,d\mathbf{x}'$$
$$-\frac{e^2}{2}\iint \frac{n(\mathbf{x}, \mathbf{x}')n(\mathbf{x}', \mathbf{x})}{|\mathbf{x} - \mathbf{x}'|}\,d\mathbf{x}\,d\mathbf{x}' + \sum_{i=2} e^{2i}\,\Gamma_i[n], \tag{6.56}$$

coincides with the decomposition proposed by Kohn and Sham [9]. Therefore the inversion method of evaluating the effective action functional [13] naturally leads to the Kohn–Sham density-functional theory. Application of the variational principle (see Equation (6.17)) to the expansion (6.56) yields the well-known Kohn–Sham self-consistent procedure and the corresponding single-particle equations. Comparison of Equations (6.3) and (6.56) immediately provides an expression for the exchange-correlation functional. These topics are discussed in detail in the following sections.

To conclude this section, we would like to note that alternatively, the effective action functional can be also evaluated via the auxiliary field method [4] However, in this method the Kohn–Sham decomposition has to be artificially imposed to allow the study of the exchange-correlation functional [24] or Kohn–Sham density-functional theory in general.

6.4.2.   Construction of the exchange-correlation functional

The success of first-principles calculations based on Kohn–Sham density-functional theory depends on the accuracy of the approximations to the exchange-correlation functional. The analysis of Kohn–Sham theory, or E_{xc} in particular, via standard many-body perturbation theory [11,12] was always a challenging task, for there was no explicit connection between the two methods. The advantage of the effective action formalism is that it is a rigorous many-body approach specifically designed for a density-based description of many-body systems. This formalism provides a natural definition of the exchange-correlation functional as,

$$E_{xc}[n] = -\frac{1}{2\beta}\left(\,\bigcirc\,\right) + \sum_{i=2} e^{2i}\,\Gamma_i[n].$$ (6.57)

This expression involves only Kohn–Sham based quantities and is especially suitable for practical applications. Simple diagrammatic rules for evaluating higher order terms in the expansion of $\Gamma[n]$ are readily available [5,13]. This, in turn leads to the following set of rules for the calculation of the exchange-correlation functional:

1. Draw all connected diagrams made of Kohn–Sham propagators $\mathcal{G}_0(x, x')$ and Coulomb interaction lines $u(x - x')$ with the corresponding weight factors [25].
2. Eliminate all the graphs that can be separated by cutting a single Coulomb interaction line.
3. For each two-particle reducible (2PR) graph (i.e., any graph that can be separated by cutting two propagator lines) perform the following procedure [5].
   (a) Separate the graph by cutting 2PR propagators.
   (b) For each of the two resulting graphs join two external propagators.
   (c) Connect the two graphs via the inverse density propagator $\mathcal{D}(\mathbf{x}, \mathbf{x}')$.
   (d) Repeat the procedure until no new graph is produced.
   (e) Sum up all the resulting graphs including the original graph.

The inverse density propagator $\mathcal{D}(\mathbf{x}, \mathbf{x}')$ is given by,

$$\mathcal{D}(\mathbf{x}, \mathbf{x}') = -\left[\int_0^\beta \mathcal{G}_0(\mathbf{x}\tau, \mathbf{x}'\tau')\mathcal{G}_0(\mathbf{x}'\tau', \mathbf{x}\tau)\,d\tau'\right]^{-1},$$ (6.58)

or in terms of Kohn–Sham orbitals

$$\mathcal{D}(\mathbf{x}, \mathbf{x}') = -\left[\sum_{i\neq j}(n_i - n_j)\frac{\varphi_i(\mathbf{x})\varphi_i^*(\mathbf{x}')\varphi_j(\mathbf{x}')\varphi_j^*(\mathbf{x})}{\varepsilon_i - \varepsilon_j}\right]^{-1}.$$ (6.59)

Application of these rules for generation of the first-order correction to $E_{xc}[n]$ is obvious and leads to the well-known Kohn–Sham exchange functional [14,15,24,26].

$$E_{xc,1} = -\frac{1}{2\beta}\left(\,\bigcirc\,\right).$$ (6.60)

Let us now consider the second order correction to $E_{xc}[n]$. Rule #1 leads to the already given expression for $W_2[J_0]$ (see Equation (6.50)). Application of Rule #2 leads to elimination of the last two graphs, thus giving

$$\frac{1}{4\beta}\left(\,\otimes\,\right) - \frac{1}{4\beta}\left(\,\bigcirc\!\cdots\!\bigcirc\,\right) + \frac{1}{2\beta}\left(\,\bigcirc\,\right).$$ (6.61)

The third graph in the above expression is 2PR. According to Rule #3 it transforms to:

$$\bigcirc\!\!\!\!\bigcirc \Rightarrow \bigcirc\!\!\!\!\bigcirc + \bigcirc\!\!-\!\!\bigcirc. \tag{6.62}$$

Here double solid line denotes the inverse density propagator $\mathcal{D}(\mathbf{x}, \mathbf{x}')$. Therefore the final expression for the second order correction to E_{xc} is given by,

$$E_{xc,2} = \frac{1}{4\beta}\,\bigotimes - \frac{1}{4\beta}\,\bigcirc\!\!\cdots\!\!\bigcirc$$

$$+ \frac{1}{2\beta}\,\bigcirc + \frac{1}{2\beta}\,\bigcirc\!\!=\!\!\bigcirc. \tag{6.63}$$

It is instructive to apply the above procedure to the case of a homogeneous electron gas in the zero temperature limit. It can be demonstrated that in this case Rule #3 leads to the complete elimination of 2PR graphs. Indeed, at this limit one can show that

$$\bullet\!\!\bullet\!\!\bullet + \bullet\!\!\bigcirc\!\!-\!\!\bullet\!\!\bullet = 0. \tag{6.64}$$

Here, black circles denote parts of the diagram that are connected to each other via two propagators. For example, in the expression for the second order correction to E_{xc} the last two graphs completely cancel each other

$$\bigcirc + \bigcirc\!\!-\!\!\bigcirc = 0, \tag{6.65}$$

and as expected [11],

$$E_{xc,2}^{\text{hom}} = \frac{1}{4\beta}\,\bigotimes - \frac{1}{4\beta}\,\bigcirc\!\!\cdots\!\!\bigcirc. \tag{6.66}$$

Local Density Approximation (LDA) represents a popular choice for $E_{xc}[n]$ in first-principles calculations. This approximation and subsequent corrections can obtained via the derivative expansion [27,28] of the exchange-correlation functional

$$E_{xc}[n] = \int \left(E_{xc}^{(0)}\big(n(\mathbf{x})\big) + E_{xc}^{(2)}\big(n(\mathbf{x})\big)\big(\nabla n(\mathbf{x})\big)^2 + \cdots \right) d\mathbf{x} \tag{6.67}$$

where $E_{xc}^{(k)}$ is a *function* of $n(\mathbf{x})$, not a functional.

LDA corresponds to the first term in this expansion;

$$E_{xc}^{\text{LDA}}[n] \equiv \int E_{xc}^{(0)}\left(n(\mathbf{x})\right) d\mathbf{x}. \tag{6.68}$$

Function $E_{xc}^{(0)}(n(\mathbf{x}))$ can be found by evaluating the above expansion at constant density $n(\mathbf{x}) = n_0$

$$E_{xc}^{(0)}(n_0) = \frac{1}{V} E_{xc}[n_0], \quad n_0 = const. \tag{6.69}$$

Here $E_{xc}[n_0]$ represents the exchange-correlation energy of the homogeneous electron gas with density n_0.

Regarding the results obtained in this section, we would like to emphasize the following points. Not only does the effective action formalism leads to a straightforward set of rules to calculate the exchange-correlation functional up to any arbitrary order, but it can also be used to generate similar quantities for descriptions based on the observables other than particle density (for example, current-density functional theory). One should also remember that the expansion of $E_{xc}[n]$ represents only part of the general picture provided by the effective action formalism. As we show later, the same formalism allows us to develop a rigorous and systematic Kohn–Sham theory for one-electron propagators and many-body excitation energies.

6.4.3. Kohn–Sham self-consistent procedure

As we have demonstrated earlier, application of the variational principle leads to the Kohn–Sham self-consistent procedure. In the case of the traditional density-functional theory, the typical self-consistent procedure takes the form

1. Start with some reasonable guess for the Kohn–Sham potential $J_0(\mathbf{x})$.
2. Solve Kohn–Sham single-particle equations.
3. Determine new Kohn–Sham potential $J_0(\mathbf{x})$ using

$$J_0(\mathbf{x}) = \int \frac{n(\mathbf{x}')}{|\mathbf{x} - \mathbf{x}'|} d\mathbf{x}' + \upsilon_{xc}(\mathbf{x}), \tag{6.70}$$

where the exchange-correlation potential $\upsilon_{xc}(\mathbf{x})$ is defined as,

$$\upsilon_{xc}(\mathbf{x}) = \frac{\delta E_{xc}[n]}{\delta n(\mathbf{x})}. \tag{6.71}$$

4. Repeat from step 2 until self-consistency is achieved.

In LDA the exchange-correlation functional is an explicit functional of electron density $n(\mathbf{x})$, and the exchange-correlation potential $\upsilon_{xc}(\mathbf{x})$ can be obtained by straightforward differentiation. In the case of the diagrammatic expansion, this simple property no longer holds and $E_{xc}[n]$ appears as an implicit functional of $n(\mathbf{x})$. The exchange-correlation potential $\upsilon_{xc}(\mathbf{x})$ can still be found using the equation:

$$\upsilon_{xc}(\mathbf{x}) = -\int \mathcal{D}(\mathbf{x}, \mathbf{x}') \frac{\delta E_{xc}[n]}{\delta J_0(\mathbf{x}')} d\mathbf{x}', \tag{6.72}$$

The functional derivative in the above expression can be easily evaluated based on the following relationships:

$$\frac{\delta \mathcal{G}_0(x_1, x_2)}{\delta J_0(\mathbf{x})} = \int_0^\beta \mathcal{G}_0(x_1, x)\mathcal{G}_0(x, x_2)\, d\tau, \tag{6.73}$$

and

$$\frac{\delta \mathcal{D}(\mathbf{x}, \mathbf{x}')}{\delta \mathcal{G}_0(x_1, x_2)} = 2\mathcal{D}(\mathbf{x}, \mathbf{x}_1)\mathcal{G}_0(x_1, x_2)\mathcal{D}(\mathbf{x}_2, \mathbf{x}'). \tag{6.74}$$

For example, the first order correction to $v_{xc}(\mathbf{x})$ is given by,

$$v_{xc,1}(\mathbf{x}) = \frac{1}{\beta} \quad \text{} \tag{6.75}$$

Obviously, even in case of the diagrammatic expansion of E_{xc} one could still use the above mentioned self-consistent procedure. To reduce the computational effort, however, slight modification of that procedure might be advantageous. Namely, we suggest to shift the emphasis from density n to Kohn–Sham potential J_0. Indeed, Corollary 1 guarantees that there is a one-to-one correspondence between n and J_0. Thus, we can consider the effective action functional that depends on J_0 rather than n:

$$\overline{\Gamma}[J_0] \equiv \Gamma\big[n[J_0]\big]. \tag{6.76}$$

The variational principle then takes the form

$$\frac{\delta \overline{\Gamma}[J_0]}{\delta J_0(\mathbf{x})} = 0. \tag{6.77}$$

In other words, one has to find a Kohn–Sham potential that minimizes the effective action functional $\overline{\Gamma}[J_0]$. To accomplish this task, one could use the so-called steepest descent minimization method [29]. In this case the self-consistent procedure takes the form:

1. Start with some reasonable guess for the Kohn–Sham potential $J_0(\mathbf{x})$.
2. Calculate the direction of steepest descent $s(\mathbf{x})$ as,

$$s(\mathbf{x}) = -\frac{\delta \overline{\Gamma}[J_0]}{\delta J_0(\mathbf{x})} \tag{6.78}$$

   or

$$s(\mathbf{x}) = -\int \mathcal{D}^{-1}(\mathbf{x}, \mathbf{x}') J_0(\mathbf{x}')\, d\mathbf{x}' - \frac{\delta \overline{\Gamma}_{int}[J_0]}{\delta J_0(\mathbf{x})}. \tag{6.79}$$

3. Determine new Kohn–Sham potential $J_0^{new}(\mathbf{x})$ from the old Kohn–Sham potential $J_0^{old}(\mathbf{x})$ by stepping along the direction of steepest descent:

$$J_0^{new}(\mathbf{x}) = J_0^{old}(\mathbf{x}) + \alpha s(\mathbf{x}) \tag{6.80}$$

   where α is the length of the step.
4. Repeat from step 2 until self-consistency is achieved.

The advantage of this self-consistent procedure is that it is much easier to calculate

$$\frac{\delta \overline{\Gamma}[J_0]}{\delta J_0(\mathbf{x})}$$

rather than

$$\frac{\delta \Gamma[n]}{\delta n(\mathbf{x})}.$$

For example, when $\overline{\Gamma}_{int}[J_0]$ is approximated by its first order correction

$$\overline{\Gamma}_{int}[J_0] \approx \frac{1}{2\beta}\;\bigcirc\!\cdots\!\bigcirc \;-\; \frac{1}{2\beta}\;\ominus \tag{6.81}$$

the steepest descent direction is given by

$$s(\mathbf{x}) = -\int \mathcal{D}^{-1}(\mathbf{x}, \mathbf{x}')\left(J_0(\mathbf{x}') - \int \frac{n(\mathbf{x}'')}{|\mathbf{x}'' - \mathbf{x}'|}\,d\mathbf{x}''\right)d\mathbf{x}'$$

$$+ \int_0^\beta d\tau \iint \mathcal{G}_0(\mathbf{y}0, \mathbf{x}\tau)\mathcal{G}_0(\mathbf{x}\tau, \mathbf{y}'0)\frac{n(\mathbf{y}', \mathbf{y})}{|\mathbf{y}' - \mathbf{y}|}\,d\mathbf{y}\,d\mathbf{y}'. \tag{6.82}$$

The advantage of this expression as compared to the first order correction to $v_{xc,1}$ is that here, we avoid the calculation of the inverse density propagator $\mathcal{D}(\mathbf{x}, \mathbf{x}')$. Note that $\mathcal{D}^{-1}(\mathbf{x}, \mathbf{x}')$ can easily be written in terms of Kohn–Sham single-particle orbitals and energies (see Equation (6.59)).

6.5. Time-dependent probe

To study excitation energies and one-electron propagators, it is necessary to consider an imaginary-time-dependent probe. The definition of the functional $W[J]$ is changed correspondingly

$$e^{-W[J]} = \int D\psi^\dagger D\psi\, e^{-S[\psi^\dagger, \psi] - \int J(x)\psi^\dagger(x)\psi(x)\,dx}. \tag{6.83}$$

Note that the parameter β has been absorbed into $W[J]$. In order to proceed with the inversion method, we need to assure that the map $J(x) \to n(x)$,

$$n(x) = \frac{\delta W[J]}{\delta J(x)}, \tag{6.84}$$

is invertible. Since at the end the time-dependent probe is set to zero, we can assume our source to be infinitesimally small. Therefore it suffices to prove the invertibility in the small neighborhood of *time-independent* external source:

$$J(\mathbf{x}) + \delta J(\mathbf{x}) \to n(\mathbf{x}) + \delta n(x). \tag{6.85}$$

Since,

$$\delta n(x) = \int \left(\frac{\delta^2 W[J]}{\delta J(x)\delta J(x')} \right)_{J(\mathbf{x})} \delta J(x')\, dx', \tag{6.86}$$

we need to show that the operator

$$W^{(2)}(x, x') \equiv \left(\frac{\delta^2 W[J]}{\delta J(x)\delta J(x')} \right)_{J(\mathbf{x})} \tag{6.87}$$

has no zero eigenvalues. This property follows from the following theorem.

THEOREM 2.   *The operator $W^{(2)}(x, x')$ is strictly negative definite.*

(The proof is given in the at the end of this chapter.)
This theorem guarantees that as long as two infinitesimally small probes differ by more than a pure time-dependent function, they would produce two different densities. Once this one-to-one correspondence has been established, further analysis proceeds similar to the time-independent case. The effective action functional is defined as,

$$\Gamma[n] = W[J] - \int J(x)n(x)\, dx, \tag{6.88}$$

where J is assumed to be a functional of n by Equation (6.84). Using the inversion method [13], the effective action functional can be found as a power series in terms of the coupling constant:

$$\Gamma\left[n, e^2\right] = \Gamma_0[n] + e^2\Gamma_1[n] + e^4\Gamma_2[n] + \cdots . \tag{6.89}$$

Zeroth order term is given by,

$$\Gamma_0[n] = W_0[J_0] - \int J_0(x)n(x)\, dx. \tag{6.90}$$

The functional $W_0[J_0]$ describes a Kohn–Sham system of noninteracting electrons in the presence of an imaginary-time-dependent external potential $J_0(x)$. Kohn–Sham potential $J_0(x)$ is chosen such that the time-dependent density $n(x)$ is reproduced. Diagrammatic structure of $\Gamma_i[n]$ is the same as its time-independent counterparts with the only difference being that Kohn–Sham propagator is now defined in the presence of a time-dependent Kohn–Sham potential $J_0(x)$,

$$\mathcal{G}_0^{-1}(x, x') = -\left(\frac{\partial}{\partial \tau} - \frac{\nabla^2}{2m} + v_{ion}(\mathbf{x}) + J_0(x) \right)\delta(x - x'), \tag{6.91}$$

and the inverse density propagator becomes

$$\mathcal{D}(x, x') = -\left[\mathcal{G}_0(x, x')\mathcal{G}_0(x', x)\right]^{-1}. \tag{6.92}$$

6.6. One-electron propagators

Using the effective action formalism it is possible to develop a systematic Kohn–Sham density-functional approach to one-electron propagators. It is common practice to use converged Kohn–Sham single-particle orbitals and energies in quasiparticle calculations. However, very often these methods are not very systematic in their use of Kohn–Sham based quantities [19]. The formalism presented below provides a rigorous theoretical foundation for the calculation of quasiparticle properties based on Kohn–Sham noninteracting system.

Consider $W[J]$ in the presence of the auxiliary nonlocal source $\xi(x, x')$

$$e^{-W[J]} = \int D\psi^\dagger D\psi \, e^{-S_\xi[\psi^\dagger, \psi]}, \tag{6.93}$$

where

$$S_\xi[\psi^\dagger, \psi] = S[\psi^\dagger, \psi] + \int J(x)\psi^\dagger(x)\psi(x) \, dx$$
$$+ \iint \xi(x, x')\psi^\dagger(x)\psi(x') \, dx \, dx'. \tag{6.94}$$

The nonlocal source, $\xi(x, x')$, allows us to write the one-electron propagator or the (finite temperature) Green's function, $\mathcal{G}(x, x') = -\langle T_\tau \psi(x)\psi^\dagger(x')\rangle$, as a functional derivative

$$\mathcal{G}(x, x') = \left(\frac{\delta W[J]}{\delta \xi(x', x)}\right)_J. \tag{6.95}$$

Using the well-known property of the Legendre transformation [30],

$$\left(\frac{\delta W[J]}{\delta \xi(x', x)}\right)_J = \left(\frac{\delta \Gamma[n]}{\delta \xi(x', x)}\right)_n, \tag{6.96}$$

the one-electron propagator can be expressed in terms of the effective action functional as,

$$\mathcal{G}(x, x') = \left(\frac{\delta \Gamma_0[n]}{\delta \xi(x', x)}\right)_n + \left(\frac{\delta \Gamma_{int}[n]}{\delta \xi(x', x)}\right)_n. \tag{6.97}$$

Using the property (6.96) we obtain that

$$\left(\frac{\delta \Gamma_0[n]}{\delta \xi(x', x)}\right)_n = \left(\frac{\delta W_0[J_0]}{\delta \xi(x', x)}\right)_{J_0} = \mathcal{G}_0(x, x'). \tag{6.98}$$

Let us consider the second term in Equation (6.97). One can show that

$$\left(\frac{\delta \Gamma_{int}[n]}{\delta \xi(x', x)}\right)_n = \iint \frac{\delta \Gamma_{int}[n]}{\delta \mathcal{G}_0(y, y')} \mathcal{G}_0(y, x')\mathcal{G}_0(x, y') \, dy \, dy'$$
$$- \int \frac{\delta \Gamma_{int}[n]}{\delta n(y)} \mathcal{G}_0(y, x')\mathcal{G}_0(x, y) \, dy. \tag{6.99}$$

Our original system is recovered by setting ξ to zero. Using Equations (6.97), (6.98), (6.99) the exact one-electron propagator (in operator notation) is given by,

$$\mathcal{G} = \mathcal{G}_0 + \mathcal{G}_0 \cdot \Sigma_0 \cdot \mathcal{G}_0, \tag{6.100}$$

where the Kohn–Sham self-energy Σ_0 is given by,

$$\Sigma_0(x_1, x_2) = \frac{\delta \Gamma_{int}[n]}{\delta \mathcal{G}_0(x_2, x_1)} - J_0(x_1)\delta(x_1 - x_2) \qquad (6.101)$$

and $J_0(x_1)$ is the Kohn–Sham potential,

$$J_0(x) = \frac{\delta \Gamma_{int}[n]}{\delta n(x)}. \qquad (6.102)$$

The functional derivative in Equation (6.101) can be easily evaluated since the functional Γ_{int} can be expressed entirely in terms of Kohn–Sham Green's functions $\mathcal{G}_0$. The above expression for the one-electron propagator can also be written as,

$$\mathcal{G} = \mathcal{G}_0 + \mathcal{G}_0 \cdot \tilde{\Sigma}_0 \cdot \mathcal{G}, \qquad (6.103)$$

where

$$\tilde{\Sigma}_0 = \Sigma_0 \cdot (1 + \mathcal{G}_0 \cdot \Sigma_0)^{-1} = \left(\Sigma_0^{-1} + \mathcal{G}_0\right)^{-1}.$$

Therefore the exact self-energy Σ is given by,

$$\Sigma(x, x') = J_0(x)\delta(x - x') + \tilde{\Sigma}_0(x, x'). \qquad (6.104)$$

The above formulation provides a systematic way to study one-electron propagators and self-energy in Kohn–Sham density-functional theory. The important feature of this formulation is that both self-energy and Kohn–Sham potential are determined from one quantity $\Gamma_{int}[n]$. *A single approximation to the functional $\Gamma_{int}[n]$ simultaneously generates both the self-energy and Kohn–Sham potential.* This is to be contrasted with the common strategy of performing *separate* approximations for Kohn–Sham potential and the self-energy [19].

6.7.   Excitation energies

In addition to one-electron propagators, the effective action formalism allows a systematic study of the many-body excited states in density-functional theory [4]. Consider the Fourier transform of $W^{(2)}(x, x')$:

$$W^{(2)}(\mathbf{x}, \mathbf{x}', i\nu_s) = \int_0^\beta W^{(2)}(x, x')e^{i\nu_s(\tau - \tau')}\, d(\tau - \tau'), \qquad (6.106)$$

where

$$\nu_s = 2\pi s/\beta, \qquad (6.107)$$

It can be analytically continued into the complex plane ω:

$$W^{(2)}(\mathbf{x}, \mathbf{x}', \omega) = W^{(2)}(\mathbf{x}, \mathbf{x}', i\nu_s)|_{i\nu_s \to \omega + i\eta}. \qquad (6.108)$$

The proposition below guarantees that this analytic continuation has an inverse in the upper complex plane ω including the real axis.

PROPOSITION 2. *The operator* $W^{(2)}(\mathbf{x}, \mathbf{x}', \omega)$ *has no zero eigenvalues when ω is located in the upper half of the complex plane including the real axis.*

(See the end of this chapter for a proof.)

Let us define the excitation kernel as,

$$\Gamma^{(2)}(x, x') = \left(\frac{\delta \Gamma^2[n]}{\delta n(x) \delta n(x')} \right)_{n(\mathbf{x})}. \tag{6.109}$$

It easy to show that

$$\int W^{(2)}(x, x') \Gamma^{(2)}(x', y) \, dx' = -\delta(x - y), \tag{6.110}$$

or in terms of Fourier transforms

$$\int W^{(2)}(\mathbf{x}, \mathbf{x}', i v_s) \Gamma^{(2)}(\mathbf{x}', \mathbf{y}, -i v_s) \, d\mathbf{x}' = -\delta(\mathbf{x} - \mathbf{y}). \tag{6.111}$$

PROPOSITION 3. $\Gamma^{(2)}(\mathbf{x}, \mathbf{y}, -i v_s)$ *has a unique analytic continuation* $\Gamma^{(2)}(\mathbf{x}, \mathbf{y}, -\omega)$ *such that*

(1) *it does not have any zeros in the upper half of the complex plane ω including the real axis.*

(2) $[\Gamma^{(2)}(\mathbf{x}, \mathbf{y}, -\omega)]^{-1} \sim \frac{1}{|\omega|}$ $(\omega \to \infty)$.

The proof is given at the end of this chapter. Based on the above statement we can extend the relationship (6.111) to the whole complex plane ω:

$$\int W^{(2)}(\mathbf{x}, \mathbf{x}', \omega) \Gamma^{(2)}(\mathbf{x}', \mathbf{y}, -\omega) \, d\mathbf{x}' = -\delta(\mathbf{x} - \mathbf{y}). \tag{6.112}$$

Consider $W^{(2)}(\mathbf{x}, \mathbf{x}', \omega)$ in the zero temperature limit $(\beta \to \infty)$. It is well known that it has poles just below the real axis at the exact excitation energies,

$$\omega = E_l - E_0 - i\eta. \tag{6.113}$$

Then it follows from Equation (6.112) that $\Gamma^{(2)}(\mathbf{x}, \mathbf{y}, -\omega)$ has zero eigenvalues at the exact excitation energies [4]. In other words, when $\omega = E_l - E_0 - i\eta$, there exists $\xi(\mathbf{x})$ such that [4]

$$\int \Gamma^{(2)}(\mathbf{x}, \mathbf{y}, -\omega) \xi(\mathbf{x}) \, d\mathbf{x} = 0. \tag{6.114}$$

Using the coupling constant expansion for $\Gamma(n)$ the excitation kernel becomes

$$\Gamma^{(2)}(\mathbf{x}, \mathbf{y}, -i v_s) = \Gamma_0^{(2)}(\mathbf{x}, \mathbf{y}, -i v_s) + \Gamma_{int}^{(2)}(\mathbf{x}, \mathbf{y}, -i v_s). \tag{6.115}$$

Substituting this in Equation (6.114) and using the fact that,

$$\left[\Gamma_0^{(2)}(\mathbf{x}, \mathbf{x}', -\omega) \right]^{-1} = -W_0^{(2)}(\mathbf{x}, \mathbf{x}', \omega), \tag{6.116}$$

we obtain

$$\xi(\mathbf{x}) = \iint W_0^{(2)}(\mathbf{x}, \mathbf{x}', \omega) \Gamma_{int}^{(2)}(\mathbf{x}', \mathbf{y}', -\omega) \xi(\mathbf{y}') \, d\mathbf{x}' d\mathbf{y}'. \tag{6.117}$$

Here $W_0^{(2)}(\mathbf{x}, \mathbf{x}', \omega)$ represents the negative of the density-correlation function for Kohn–Sham noninteracting system;

$$W_0^{(2)}(\mathbf{x}, \mathbf{x}', \omega) = \sum_l \frac{n_l^*(\mathbf{x})n_l(\mathbf{x}')}{\omega - \omega_l^{ks} + i\eta} - \frac{n_l(\mathbf{x})n_l^*(\mathbf{x}')}{\omega + \omega_l^{ks} + i\eta}, \tag{6.118}$$

where

$$\langle l^{ks} | \hat{n}(\mathbf{x}) | 0^{ks} \rangle = n_l(\mathbf{x}), \tag{6.119}$$

and

$$\omega_l^{ks} = E_l^{ks} - E_0^{ks}. \tag{6.120}$$

Here all the quantities refer to the Kohn–Sham noninteracting system. Aside from the notational differences, Equation (6.117) coincides with similar expressions derived using time-dependent DFT [20,21]. Searching for a solution of the form

$$\xi(\mathbf{x}) = \sum a_l n_l(\mathbf{x}) + \sum b_l n_l^*(\mathbf{x}), \tag{6.121}$$

we obtain the following matrix equation [21],

$$\begin{bmatrix} L & M \\ M^* & L^* \end{bmatrix}\begin{bmatrix} A \\ B \end{bmatrix} = (\omega + i\eta)\begin{bmatrix} -1 & 0 \\ 0 & 1 \end{bmatrix}\begin{bmatrix} A \\ B \end{bmatrix}, \tag{6.122}$$

where

$$M_{ij} = \iint n_i^*(\mathbf{x})\Gamma_{int}^{(2)}(\mathbf{x}, \mathbf{y}, -\omega)n_j^*(\mathbf{y})\,d\mathbf{x}\,d\mathbf{y}, \tag{6.123}$$

$$L_{ij} = \iint n_i^*(\mathbf{x})\Gamma_{int}^{(2)}(\mathbf{x}, \mathbf{y}, -\omega)n_j(\mathbf{y})\,d\mathbf{x}\,d\mathbf{y} \tag{6.124}$$

$$+ (E_i - E_0)\delta_{ij}, \tag{6.125}$$

and

$$(A)_i = a_i, \qquad (B)_i = b_i. \tag{6.126}$$

The values of ω for which the above matrix equation has nontrivial solutions, is determined by the condition

$$\det\left(\begin{bmatrix} L & M \\ M^* & L^* \end{bmatrix} - (\omega + i\eta)\begin{bmatrix} -1 & 0 \\ 0 & 1 \end{bmatrix}\right) = 0. \tag{6.127}$$

Therefore, the effective action formalism presents an alternative way (as compared to time-dependent density-functional theory) for calculating the excitation energies. However, in addition, the effective action formalism also provides a means of calculating the exchange-correlation kernel. This feature is missing in conventional time-dependent density-functional theory [22].

6.8. Theorems involving functionals $W[J]$ and $\Gamma[Q]$

6.8.1. Time-independent probe

THEOREM 1. *The functional* $W[J] = -\frac{1}{\beta}\ln Tr(e^{-\beta(\hat{H}+J(1)\hat{Q}(1))})$ *is strictly concave, i.e., for any* α, $0 < \alpha < 1$, *and* $J \neq J'$

$$W\big[\alpha J + (1-\alpha)J'\big] > \alpha W[J] + (1-\alpha)W[J']. \tag{6.128}$$

PROOF. (Based on Ref. [31].) Consider two Hermitian operators $\hat{A}$ and $\hat{B}$. Let $\{\Psi_i\}$ be a complete set of eigenstates of the operator $\alpha\hat{A} + (1-\alpha)\hat{B}$. Then,

$$Tr\big(e^{\alpha\hat{A}+(1-\alpha)\hat{B}}\big) = \sum_i \langle\Psi_i|e^{\alpha\hat{A}+(1-\alpha)\hat{B}}|\Psi_i\rangle$$

$$= \sum_i e^{\alpha\langle\Psi_i|\hat{A}|\Psi_i\rangle + (1-\alpha)\langle\Psi_i|\hat{B}|\Psi_i\rangle}. \tag{6.129}$$

Using Hölder's inequality [32] ($a_i, b_i \geqslant 0$),

$$\sum_i a_i^{\alpha} b_i^{1-\alpha} \leqslant \left(\sum_i a_i\right)^{\alpha}\left(\sum_i b_i\right)^{1-\alpha}, \tag{6.130}$$

we obtain that

$$Tr\big(e^{\alpha\hat{A}+t(1-\alpha)\hat{B}}\big) \leqslant \left(\sum_i e^{\langle\Psi_i|\hat{A}|\Psi_i\rangle}\right)^{\alpha}\left(\sum_i e^{\langle\Psi_i|\hat{B}|\Psi_i\rangle}\right)^{(1-\alpha)}. \tag{6.131}$$

From Hölder's inequality it follows that the equality in the above expression holds only if for any i

$$\langle\Psi_i|\hat{A}|\Psi_i\rangle = \langle\Psi_i|\hat{B}|\Psi_i\rangle + \chi, \tag{6.132}$$

where χ is constant independent of i.

Since e^x is a convex function,

$$\sum_i e^{\langle\Psi_i|\hat{A}|\Psi_i\rangle} \leqslant \sum_i \langle\Psi_i|e^{\hat{A}}|\Psi_i\rangle,$$

$$\sum_i e^{\langle\Psi_i|\hat{B}|\Psi_i\rangle} \leqslant \sum_i \langle\Psi_i|e^{\hat{B}}|\Psi_i\rangle. \tag{6.133}$$

Equal sign in the above equations holds only if $\{\Psi_i\}$ are the eigenstates $\hat{A}$ and $\hat{B}$. Collecting all the results together we obtain the following inequality:

$$Tr\big(e^{\alpha\hat{A}+(1-\alpha)\hat{B}}\big) \leqslant \big(Tr\, e^{\hat{A}}\big)^{\alpha}\big(Tr\, e^{\hat{B}}\big)^{(1-\alpha)}. \tag{6.134}$$

Equality holds here if and only if operators $\hat{A}$ and $\hat{B}$ differ by a constant:

$$\hat{A} = \hat{B} + const. \tag{6.135}$$

Indeed:

$\Rightarrow$) Suppose Equation (6.135) is true, then equal sign in (6.134) is obvious.

$\Leftarrow$) If there is an equal sign in (6.134) then operators $\hat{A}$ and $\hat{B}$ must have a common set of eigenstates $\{\Psi_i\}$ and condition (6.132) must be true. Then considering the representation of operators $\hat{A}$ and $\hat{B}$ in the basis of eigenfunctions $\{\Psi_i\}$ we obtain

$$\hat{A} = \hat{B} + const. \tag{6.136}$$

Therefore

$$Tr\left(e^{\alpha \hat{A} + (1-\alpha)\hat{B}}\right) < \left(Tr\, e^{\hat{A}}\right)^{\alpha}\left(Tr\, e^{\hat{B}}\right)^{(1-\alpha)}, \tag{6.137}$$

when $\hat{A} \neq \hat{B} + const.$

Setting

$$\hat{A} = -\beta\left(\hat{H} + J(1)\hat{Q}(1)\right) \tag{6.138}$$

and

$$\hat{B} = -\beta\left(\hat{H} + J'(1)\hat{Q}(1)\right), \tag{6.139}$$

we can easily obtain

$$W\left[\alpha J + (1-\alpha)J'\right] > \alpha W[J] + (1-\alpha)W[J'], \tag{6.140}$$

when $J \neq J'$. $\qquad\qquad\qquad\qquad\qquad\qquad\qquad\qquad\qquad\qquad\qquad\qquad\square$

COROLLARY 1. *The map $\mathcal{J} \to \mathcal{Q}$ is one-to-one.*

PROOF. Consider the functional

$$\Lambda[J] = W[J] - J(1')Q(1'), \tag{6.141}$$

where J and Q are considered to be independent. Since $W[J]$ is strictly concave (Theorem 1), it follows that $\Lambda[J]$ is strictly concave. Therefore if $\Lambda[J]$ has an extremum it is unique. Hence if the equation

$$\frac{\delta \Lambda[J]}{\delta J(1)} = \frac{\delta W[J]}{\delta J(1)} - Q(1) = 0 \tag{6.142}$$

has a solution, it is unique. $\qquad\qquad\qquad\qquad\qquad\qquad\qquad\qquad\qquad\qquad\qquad\square$

PROPOSITION 1. *The effective action functional $\Gamma[Q]$ defined on the set $\mathcal{Q}$ is strictly convex.*

PROOF. Consider the family of the functionals $\{\Lambda[J, Q]: J \in \mathcal{J}, Q \in \mathcal{Q}\}$,

$$\Lambda[J, Q] = W[J] - J(1')Q(1'). \tag{6.143}$$

Here Q is considered to be independent of J. Obviously $\Lambda[J, Q]$ is linear in Q. The effective action functional can be defined as

$$\Gamma[Q] = \sup\{\Lambda[J, Q], J \in \mathcal{J}\}. \tag{6.144}$$

The above expression represents the most general way to define the effective action functional. In our case two definitions, Equation (6.10) and Equation (6.144), are equivalent. Convexity of $\Gamma[Q]$ follows from the fact that it is a supremum of the family of the linear functionals (in Q). Because by construction for any element $J \in \mathcal{J}$, there corresponds only one $Q \in \mathcal{Q}$, the functional $\Gamma[Q]$ is also strictly convex. □

6.8.2. *Time-dependent probe*

THEOREM 2. *The operator $W^{(2)}(x, x')$ is strictly negative definite.*

PROOF. We have to prove that there exists no function $f(\mathbf{x}, \tau) \neq \eta(\tau)$ such that

$$\iint f^*(x) W^{(2)}(x, x') f(x') \, dx \, dx' \geqslant 0. \tag{6.145}$$

The operator $W^{(2)}(x, x')$ actually represents the negative of the density correlation function in the presence of the *time-independent* source $J(\mathbf{x})$:

$$W^{(2)}(x, x') = -\frac{\int D\psi^\dagger D\psi \, \tilde{n}(x) \tilde{n}(x') e^{-S_J[\psi^\dagger, \psi]}}{\int D\psi^\dagger D\psi \, e^{-S_J[\psi^\dagger, \psi]}}, \tag{6.146}$$

where

$$S_J[\psi^\dagger, \psi] = S[\psi^\dagger, \psi] + \int J(\mathbf{x}) n(x) \, dx, \tag{6.147}$$

and $\tilde{n}(x)$ denotes density fluctuation operator,

$$\tilde{n}(x) = \psi^\dagger(x)\psi(x) - n(x). \tag{6.148}$$

Defining a Fourier transform as,

$$W^{(2)}(\mathbf{x}, \mathbf{x}', i\nu_s) = \int_0^\beta W^{(2)}(x, x') e^{i\nu_s(\tau - \tau')} \, d(\tau - \tau'),$$

$$\nu_s = 2\pi s/\beta,$$

left-hand side of (6.145) transforms into

$$\sum_{\nu_s} \iint \left(f(\mathbf{x}, \nu_s) \right)^* W^{(2)}(\mathbf{x}, \mathbf{x}', i\nu_s) f(\mathbf{x}', \nu_s) \, d\mathbf{x} \, d\mathbf{x}'. \tag{6.149}$$

Strict concavity of $W[J]$ for time-independent case guarantees that $\nu_s = 0$ term is strictly negative definite. For convenience, we ignore this term from the sum. Lehman representation for $W^{(2)}(\mathbf{x}, \mathbf{x}', i\nu_s)$ is given by,

$$W^{(2)}(\mathbf{x}, \mathbf{x}', i\nu_s) = e^{W[J]} \sum_{ml} \frac{e^{-\beta E_m} - e^{-\beta E_l}}{E_m - E_l + i\nu_s} \langle m|\tilde{n}(\mathbf{x})|l\rangle \langle l|\tilde{n}(\mathbf{x}')|m\rangle, \tag{6.150}$$

where

$$\left(\hat{H} + \int J(\mathbf{x})\hat{n}(\mathbf{x}) \right)|m\rangle = E_m|m\rangle, \tag{6.151}$$

$$\left(\hat{H} + \int J(\mathbf{x})\hat{n}(\mathbf{x})\right)|l\rangle = E_l|l\rangle. \tag{6.152}$$

Therefore an arbitrary ($v_s \neq 0$) term in the sum (6.149) can be written as,

$$e^{W[J]} \sum_{ml} \frac{e^{-\beta E_m} - e^{-\beta E_l}}{E_m - E_l + iv_s} \left| \langle m| \int \tilde{n}(\mathbf{x}) f(\mathbf{x}, v_s) \, d\mathbf{x}\, |l\rangle \right|^2, \tag{6.153}$$

or

$$e^{W[J]} \sum_{E_m < E_l} \left| \langle m| \int \tilde{n}(\mathbf{x}) f(\mathbf{x}, v_s) \, d\mathbf{x}\, |l\rangle \right|^2$$
$$\times \left(\frac{e^{-\beta E_m} - e^{-\beta E_l}}{E_m - E_l + iv_s} + \frac{e^{-\beta E_l} - e^{-\beta E_m}}{E_l - E_m + iv_s} \right). \tag{6.154}$$

After some simple algebra, we obtain

$$e^{W[J]} \sum_{E_m < E_l} 2 \left| \langle m| \int \tilde{n}(\mathbf{x}) f(\mathbf{x}, v_s) \, d\mathbf{x}\, |l\rangle \right|^2$$
$$\times \frac{e^{-\beta E_l} - e^{-\beta E_m}}{(E_l - E_m)^2 + v_s^2}(E_l - E_m). \tag{6.155}$$

Obviously, the above expression cannot be positive. Moreover, it cannot be zero since this would imply that for all $m \neq l$

$$\langle m| \int \hat{n}(\mathbf{x}) f(\mathbf{x}, v_s) \, d\mathbf{x}\, |l\rangle = 0 \tag{6.156}$$

and $\int \hat{n}(\mathbf{x}) f(\mathbf{x}, v_s) \, d\mathbf{x}$ commutes with the Hamiltonian. This is impossible unless $f(\mathbf{x}, v_s)$ is constant independent of $\mathbf{x}$ or $f(\mathbf{x}, \tau)$ is a function of τ only. This case, however, was excluded from the very beginning. Therefore, the operator $W^{(2)}(x, x')$ is strictly negative definite.   □

PROPOSITION 2. *The operator $W^{(2)}(\mathbf{x}, \mathbf{x}', \omega) = W^{(2)}(\mathbf{x}, \mathbf{x}', iv_s)|_{iv_s \to \omega + i\eta}$ has no zero eigenvalues when ω is located in the upper half of the complex plane including the real axis.*

PROOF. Suppose $f(\mathbf{x})$ is an eigenvector corresponding to a zero eigenvalue then

$$\iint f^*(\mathbf{x}) W^{(2)}(\mathbf{x}, \mathbf{x}', \omega) f(\mathbf{x}') \, d\mathbf{x} \, d\mathbf{x}' = 0. \tag{6.157}$$

Using Lehman representation we obtain

$$0 = \sum_{E_m < E_l} \left| \langle m| \int \tilde{n}(\mathbf{x}) f(\mathbf{x}) \, d\mathbf{x}\, |l\rangle \right|^2$$
$$\times \left(\frac{e^{-\beta E_m} - e^{-\beta E_l}}{z - (E_l - E_m)} - \frac{e^{-\beta E_m} - e^{-\beta E_l}}{z + (E_l - E_m)} \right)$$

where $z = \omega + i\eta$. Consider the imaginary part of the above expression,

$$0 = -4\text{Re}(z)\text{Im}(z) \sum_{E_m < E_l} \left| \langle m| \int \tilde{n}(\mathbf{x}) f(\mathbf{x}) \, d\mathbf{x} |l\rangle \right|^2$$

$$\times \frac{(E_l - E_m)(e^{-\beta E_m} - e^{-\beta E_l})}{|z^2 - (E_l - E_m)^2|^2}$$

$$\Rightarrow \text{Re}(z)\text{Im}(z) = 0. \tag{6.158}$$

By Theorem 2, the real part of z cannot be zero, since in this case $W^{(2)}(\mathbf{x}, \mathbf{x}', \omega = i\nu_s)$ is strictly negative definite and condition (6.157) cannot be satisfied. Therefore we necessarily obtain that $\text{Im}(z) = 0$ and

$$\omega + i\eta = \text{real} \quad \Rightarrow \quad \omega = \text{real} - i\eta.$$

Therefore $W^{(2)}(\mathbf{x}, \mathbf{x}', \omega)$ may have zero eigenvalues only when ω is located in the lower half of the complex plane excluding the real axis. □

PROPOSITION 3. $\Gamma^{(2)}(\mathbf{x}, \mathbf{y}, -i\nu_s)$ *has a unique analytic continuation* $\Gamma^{(2)}(\mathbf{x}, \mathbf{y}, \omega)$ *such that*

(1) *it does not have any zeros in the upper half of the complex plane* ω *including the real axis,*
(2) $[\Gamma^{(2)}(\mathbf{x}, \mathbf{y}, \omega)]^{-1} \sim \frac{1}{|\omega|}$ $(\omega \to \infty)$.

PROOF. Existence follows from the fact that $\Gamma^{(2)}(\mathbf{x}, \mathbf{y}, \omega)$ can be defined as $-[W^{(2)}(\mathbf{x}, \mathbf{y}, \omega)]^{-1}$. Suppose there exist two analytic continuations of $\Gamma^{(2)}(\mathbf{x}, \mathbf{y}, -i\nu_s)$ with the properties 1 and 2. Then $W^{(2)}(\mathbf{x}, \mathbf{x}', i\nu_s)$ will have two different analytic continuations which is impossible. □

6.9. Concluding remarks

Based on the effective action formalism [4], a rigorous formulation of a generalized Kohn–Sham theory has been derived. This formulation is specifically geared towards practical calculations of the ground and excited properties of real systems. Indeed, in the case of a particle-density based description of a nonrelativistic many-electron system, we have shown a systematic way to study the exchange-correlation functional, one-electron propagators and many-body excitation energies entirely in terms of the Kohn–Sham single-particle orbitals and energies. The presented formalism is very general and can be applied to various many-body systems for constructing Kohn–Sham like descriptions in terms of the expectation value of any general operator (e.g., spin-density, current-density-functional theory). Indeed, recent work along these line [6] has clearly shown the versatility of this approach in different physical systems.

Acknowledgements

This work is a part of the PhD thesis of M. Valiev at the University of Connecticut working with the author.

References

[1] P. Hohenberg, W. Kohn, Phys. Rev. B **136**, 864 (1964).
[2] U. von Barth, L. Hedin, J. Phys. C **5**, 1629 (1972);
    M.M. Pant, A.K. Rajagopal, Sol. State Comm. **10**, 1157 (1972).
[3] G. Vignale, M. Rasolt, Phys. Rev. Lett. **59**, 2360 (1987).
[4] R. Fukuda, T. Kotani, S. Yokojima, Prog. Theor. Phys. **92**, 833 (1994).
[5] R. Fukuda, M. Komachiya, S. Yokojima, Y. Suzuki, K. Okumura, T. Inagaki, Prog. Theor. Phys. Suppl. **121**, 428 (1995).
[6] J. Engel, Phys. Rev. C **75**, 014306 (2007).
[7] J. Polonyi, K. Sailer, Phys. Rev. B **66**, 155113 (2002).
[8] N.D. Mermin, Phys. Rev. A **137**, 1414 (1965).
[9] W. Kohn, L.J. Sham, Phys. Rev. A **140**, 1133 (1965).
[10] P.C. Hohenberg, W. Kohn, L.J. Sham, Adv. Quant. Chem. **21**, 7 (1990).
[11] R.M. Dreizler, E.K.U. Gross, Density-Functional Theory (Springer-Verlag, Berlin, Heidelberg, 1990) and references therein.
[12] R.G. Parr, W. Yang, Density-Functional Theory of Atoms and Molecules (Oxford University Press, New York, 1989).
[13] K. Okumura, Int. J. Mod. Phys. **11**, 65 (1996).
[14] L.J. Sham, Phys. Rev. B **32**, 3876 (1985).
[15] J.B. Krieger, Yan Li, G.J. Iafrate, Phys. Rev. A **45**, 101 (1992).
[16] M. Valiev, G.W. Fernando, Phys. Lett. A **227**, 265 (1997).
[17] L.J. Sham, W. Kohn, Phys. Rev. **145**, 561 (1966).
[18] M. Schlüter, L.J. Sham, Adv. Quant. Chem. **21**, 97 (1990), and references therein.
[19] R. Del Sole, L. Reining, R.W. Godby, Phys. Rev. B **49**, 8024 (1994), and references therein.
[20] M. Petersilka, U.J. Gossmann, E.K.U. Gross, Phys. Rev. Lett. **76**, 1212 (1996).
[21] R. Bauernschmitt, R. Ahlrichs, Chem. Phys. Lett. **256**, 454 (1996).
[22] E.K.U. Gross, W. Kohn, Adv. Quant. Chem. **21**, 255 (1990).
[23] The usefulness of the Legendre transformation in density-functional theory was earlier recognized by E.H. Lieb, in: R.M. Dreizler, J. da Providencia (Eds.), Density Functional Methods in Physics, Plenum Press, New York (1985);
    N. Nalewajski, J. Chem. Phys. **78**, 6112 (1983), and references therein.
[24] M. Valiev, G.W. Fernando, Phys. Rev. B **54**, 7765 (1996).
[25] J.W. Negele, H. Orland, Quantum Many-Particle Systems (Addison-Wesley, 1995).
[26] A. Görling, M. Levy, Phys. Rev. B **47**, 13105 (1993).
[27] A. Görling, M. Levy, Phys. Rev. A **50**, 196 (1994).
[28] L.H. Ryder, Quantum Field Theory (Cambridge University Press, New York, 1985).
[29] S.S. Rao, Optimization Theory and Applications (Halsted Press, New York, 1978).
[30] J. Zinn-Justin, Quantum Field Theory and Critical Phenomena (Clarendon Press, Oxford, 1993).
[31] Some of the techniques used in the proof were borrowed from R.B. Israel, Convexity in the Theory of Lattice Gases (Princeton University Press, Princeton, NJ, 1979).
[32] A.E. Taylor, General Theory of Functions and Integration (Dover Publications, Inc., New York, 1985).

Chapter 7

Magnetic Tunnel Junctions and Spin Torques

7.1. Magnetic random access memory (MRAM)

The transistor, invented more than half a century ago, completely revolutionized the technological world. In addition to the charge that is vital in the operation of a transistor, an electron also carries a spin. Spin based devices have been quietly responsible for a similar revolution in this computer age. Personal and other computers have made a tremendous impact and change in the way we live our lives since the early 1990s. World Wide Web, the information super highway, has significantly contributed to these changes. Among many other things, we get our news, do some of our shopping, communicate with others through this modern cyber-space. The fundamental principles at the heart of many of these developments are rooted in advances in basic research. It is often said that without breakthroughs in basic research, most technological advances would not have taken place. This has been well demonstrated over the entire history of civilization. In the early 1980s, some of the pioneers in computer industry were more than content with providing memory capacities under a few MB for most users. It was clearly difficult to envision what everyday application would require more memory. However, the exponential growth in these capacities over the last two decades have found more than enough uses. With such increases in memory, the hard disks have also seen their storage capabilities soaring.

There is a clear divide in today's computers between random access memory (RAM) and the hard disk storage. Hard disks are based on more permanent magnetic storage discussed earlier, and are utilized for packing bits as tightly as possible. The reading of data is done using GMR read sensors matching the bit sizes, such as those discussed in the chapter on the GMR effect. However, the read speeds for data on hard disks cannot compete with the fast CPUs of today that operate in the GHz range. Every time the system is rebooted, the computer has to transfer data from its hard disk onto its RAM. In this chapter, we will briefly examine the concept "MRAM" or magnetic RAM and how it is expected to compete with the current high speed data transfer techniques.

RAM is what is available for the computer to carry out its computing (processing) at a given instant, with fast access to data. Today, this is achieved through semiconductor technology (electronics). Imagine a grid consisting of "cells" where each cell represents a "bit" of storage. This memory element is typically a capacitor, in RAM found in the current generation of dynamic RAM (DRAM) based computers, which uses a transistor to control the current. The fabrication technology associated with these, which determines the size of such a memory element, has advanced so rapidly and has made it possible to make DRAM the highest density RAM currently available, in addition to being the least expensive. This is why it is used for the majority of RAM found in today's computers. However, since capacitors lose their charge over time, a constant power supply is necessary to refresh them. As DRAM cells become smaller, the refresh cycles get shorter and hence the need for a more continuous power supply to these cells. The memory stored as DRAM is said to be volatile with the storage being temporary or short-term; i.e., the data stored in DRAM is available for its current operations only. Also, as is well known, powering off the computer erases the data in DRAM.

Magnetic RAM (MRAM) is expected to handle various shortcomings of data transfers, providing instant boot-up capability, with high speed access, low power consumption and high density. Early MRAM devices used magnetic hysteresis in order to store information. A basic storage mechanism is tied to the magnetic state of a tunnel junction (as described in the next section). In contrast to DRAM, MRAM does not require any refreshing to maintain the magnetic state of a given bit, once it is magnetized, and hence the memory is non-volatile. Not only does this mean that it retains its memory with the power turned off, but also that there is no need for a constant power supply. While in principle the read process requires more power than in a DRAM, in reality the difference appears to be negligible. However, the write process requires more power in order to erase the magnetic state stored in the junction, varying from three to eight times the power required during reading [1]. Although the exact amount of power savings depends on the nature of the work (more frequent writing will require more power), in general MRAM proponents expect much lower power consumption (up to 99% less) compared to DRAM. Spin Transfer Torque (Ref. [2] and to be discussed later) based MRAMs are expected to eliminate the difference between reading and writing, further reducing power requirements (see Figure 7.1 for an illustration of a MRAM cell (Ref. [3])).

One other current RAM that can compete with MRAM is static RAM (SRAM), which is again based on electronics. Here several transistors are used to form a bit which makes SRAM somewhat expensive. In addition, the bit density may not be able to match the bit density of most current high density MRAM devices much longer. However, the power requirements of these transistors is very low and the switching time is very fast. Hence SRAM is sometimes used for small amounts of high speed memory such as in a CPU cache.

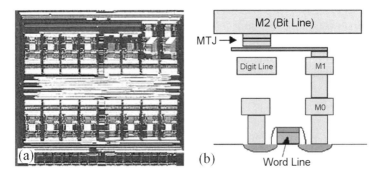

Figure 7.1.   (a) Microphotograph of fully integrated 64 Kb MRAM. (b) Schematic vertical structure of MRAM cell. Reproduced with the permission of Elsevier from Ref. [3].

7.2. Magnetic tunnel junctions

Spin-polarized tunneling, reported in 1970 by Meservey *et al.* [4], has laid the foundation to an exciting area of fundamental and applied research. Tunneling from ferromagnetic elemental metals or compounds reflect the spin polarization present in the ferromagnet. There is on-going intensive research on Magnetic Tunnel Junctions (MTJs) since these are regarded as quite promising in MRAM applications. A magnetic tunnel junction is different from the metallic sandwich structures discussed with regard to the GMR effect. A MTJ consists of two metallic layers, sometimes called electrodes, separated by an insulating layer thin enough to allow some tunneling current. The interest in such devices is at least partly due their potential applications in magnetoresistive random access memory (MRAM) devices and sensors. One of the pioneering experiments on such a system was carried out by Julliere [5] using the system Fe-Ge-Co. Figure 7.2 illustrates how the tunneling conductance decreased rapidly with the application of a bias voltage of the order of a few milli-volts in this experiment.

Conductance $G(V)$ measurements, done for the two cases where the average magnetizations in Fe and Co were parallel and antiparallel, showed a difference which could be related to the spin polarizations of the (tunneling) conduction electrons. Using a simple (stochastic) model for tunneling electrons, with a and a' representing fractions of those electrons parallel to magnetizations in Fe and Co, the Fe-Ge-Co junction conductance was expressed as $G_p \propto \{aa' + (1-a)(1-a')\}$, when magnetizations in Fe and Co are parallel (p), and $G_{ap} \propto \{a(1-a') + a'(1-a)\}$, when they are antiparallel (ap).

The relative conductance variation can be obtained (under the assumption of spin conservation) as

$$TMR = \frac{G_p - G_{ap}}{G_{ap}} = 2PP'/(1 - PP') \tag{7.1}$$

where $P = a - (1-a) = 2a - 1$ and $P' = a' - (1-a') = 2a' - 1$ are the conduction electron spin polarizations of the two ferromagnetic metals. In Julliere's original work [5]

$$JMR = \frac{G_p - G_{ap}}{G_p} = \frac{R_{ap} - R_p}{R_{ap}} = 2PP'/(1 + PP') \tag{7.2}$$

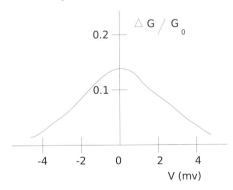

Figure 7.2. Relative conductance of Fe-Ge-Co junctions at 4.2 K. ΔG is the difference between the two conductance values corresponding to parallel and antiparallel magnetizations of the two ferromagnetic films. Reproduced with the permission of Elsevier from Ref. [5].

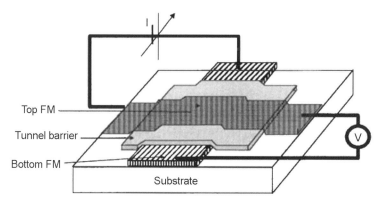

Figure 7.3.   Schematic of TMR junction geometry for 4-point measurement. Reproduced with the permission
of Elsevier from Ref. [7].

was used as a measure of the tunneling resistance change. The maximum measured value
of the relative conductance (shown in Figure 7.2) was reported to be about 14%. Spin
flips at the interface or magnetic coupling between the two ferromagnets were thought
to be responsible for any observed deviations from this simple minded expression. However, the above formalism does not contain any specific details regarding the nature of
the insulating material (amorphous Ge in this case). Although the above results were
difficult to reproduce, the phenomenon of tunneling magnetoresistance (TMR) has now
been well established in samples that are cleaner and grown carefully. However, for over
two decades after the discovery of Meservey *et al.* [4], large values of TMR were not
observed. There are many reasons for this lack of success, first and foremost being the interface roughness and the inability to switch the magnetizations of the two ferromagnets
independently. Growing thin insulating barriers on a rough surface was extremely difficult. Also, it was observed that magnetic oxide insulating barriers almost totally quenched
the TMR effect, probably affecting the spin conservation assumed in Julliere's work [5],
due to spin scattering. After addressing the above growth issues carefully, high TMR values have been obtained consistently (and reproducibly) in many laboratories around the
world since 1995.

Large values of tunneling magnetoresistance at room temperature, reported by Moodera *et al.* [6] in 1995, helped in driving the interest in TMR devices making them useful
in high-resistance, low power applications. Some other examples include tunnel barriers that use amorphous Aluminum Oxide as the insulating material. A simple schematic
illustration of a TMR junction is shown in Figure 7.3 (Ref. [7]).

In these devices, the magnetoresistance ratio (MR), defined as $(R_{ap} - R_p)/R_p =$
$(G_p - G_{ap})/G_{ap}$, has been observed to reach values as high as 70%. More recently,
in MTJs consisting of ferromagnetic amorphous CoFeB electrodes and MgO, it has been
reported that TMR at room temperature can go up to 230%. Apparently, MgO is crucial
to achieving such high TMR values.

The usual understanding of polarized currents using ferromagnetic electrodes is directly
tied to the spin polarization P of the ferromagnet, i.e.,

$$\tilde{P} = \frac{N_\uparrow - N_\downarrow}{N_\uparrow + N_\downarrow}. \tag{7.3}$$

However, it appears that the polarization of the tunneling current and (even) its sign can depend on the insulating material that is found in the MTJ. Since the mid nineties, more and more spin-dependent tunneling experiments have been conducted with oxides, such as Fe_3O_4 and CrO_2. In addition, LSMO ($La_{0.7}Sr_{0.3}MnO_3$), STO ($SrTiO_3$), and ALO (Al_2O_3) are some of the typical insulators used in MTJs and it is now well established that the polarization is an intrinsic property of the sandwich itself (including the insulator as well as the specific orientations of the ferromagnetic interfaces). In fact, the interface effects are clearly documented by the so-called "dusting experiments" on high quality samples. These have to do with adding an extremely thin Cu layer in addition to the insulating layer, which has been shown to dramatically affect the tunneling properties of the MTJ. Other developments include double barrier tunnel junctions (discussed later) and related transport properties.

7.3.   Theoretical aspects of TMR

Early theoretical attempts had resorted to numerous simplifications in order to study TMR. One such assumption is that the band structure varies slowly (compared with the electron de Broglie wavelength) in the device and the bands are parabolic, free electron type. When the barrier is low (in energy) and thin (in size), the wavefunctions from the two ferromagnets overlap in the barrier region and hence have to be matched at the interfaces. Such matching was first carried out by Slonczweski [8], who assumed parabolic but spin-split bands in the ferromagnets. He then solved the Schrödinger equation for up and down spin electrons across a simple rectangular barrier and obtained the current from the current operator, assuming that $\mathbf{k}_\parallel$ is conserved during the tunneling process.

An important conclusion of Slonczweski [8] is that the polarization of tunneling electrons, P', depends not only on the polarization of the ferromagnets, but also on the barrier height. This was expressed as

$$P' = \left[\frac{k^\uparrow - k^\downarrow}{k^\uparrow + k^\downarrow}\right]\left[\frac{\kappa^2 - k^\uparrow k^\downarrow}{\kappa^2 + k^\uparrow k^\downarrow}\right]. \tag{7.4}$$

Here $k^\uparrow$ and $k^\downarrow$ are the Fermi wave vectors for the up and down spin bands and $\hbar\kappa = \sqrt{[2m(V_b - E_F)]}$ where V_b is the barrier height and E_F is the Fermi level of the ferromagnet. From Equation (7.4), it is clear that in addition to the polarization associated with the ferromagnets, there is a barrier dependent factor which can vary from -1 to $+1$, in addition to the (Fermi level) density of states ($n^\uparrow$ and $n^\downarrow$) dependent factor

$$P = \left[\frac{k^\uparrow - k^\downarrow}{k^\uparrow + k^\downarrow}\right]\left[\frac{n^\uparrow - n^\downarrow}{n^\uparrow + n^\downarrow}\right], \tag{7.5}$$

where Equation (7.5) holds for parabolic, free electron bands.

It follows that for a strong barrier when κ is high, the polarization will differ from Julliere's result [5] and even the sign of the polarization can change. This free electron model treats the tunnel junction better that the classical approach but still, it is not that straightforward to extend this beyond parabolic bands.

There are several first principles calculations carried out during the past decade that try to address coherent tunneling in MTJs. In most of such work, ideal crystal structures are used with perfect interfaces. This is the easiest case for such a theory to investigate, since including disorder usually leads to large unit cells. Judging by the first principles work on the GMR work discussed earlier, it is important to learn the predictions of the ideal cases to gain physical insight. When the coherency of the electron wave functions is conserved, the strongly tunneling states are totally symmetric about the axis normal to the multilayers.

Recent theoretical work used the more rigorous Kubo/Landauer formula for calculating the conductance, which is outlined below. A very general result for the conductance in a given spin channel G^σ was obtained by Landauer [9], valid in the linear response (i.e., low bias) regime. This is known to be equivalent to the Kubo formula [10]. For certain special cases, such as coherent tunneling through a high barrier with N atomic planes, Landauer formula reduces to,

$$G^\sigma = \frac{4e^2}{h} \exp(-2\kappa a N)$$
$$\times \sum_{\mathbf{k}_\parallel} \frac{\mathrm{Im}(g_L{}^\sigma)\mathrm{Im}(g_R{}^\sigma)}{|1 - (g_L{}^\sigma + g_R{}^\sigma)\exp(-\kappa a) + g_L{}^\sigma g_R{}^\sigma \exp(-2\kappa a)|^2} \quad (7.6)$$

where $g_L{}^\sigma$ and $g_R{}^\sigma$ are the surface Green functions of the isolated left and right electrodes (at the Fermi energy and given $\mathbf{k}_\parallel$) and κ is related to the barrier height as before (but averaged over the 2-dimensional Brillouin zone). This tunneling conductance has a straightforward physical interpretation; the imaginary parts of the Green functions in the numerator are proportional to the 1-dimensional surface spin density of states of each channel in the left and right electrodes, which is scaled by the denominator representing the mutual interaction of the electrodes through possible overlap of the wavefunctions in the barrier region. A simple tight binding calculation of the TMR as a function of the barrier height is shown in Figure 7.4.

In early calculations of tunneling conductance, the importance of density of states was emphasized with little or no attention paid to tunneling matrix elements. A fairly comprehensive theoretical study of spin dependent tunneling conductance of Fe/MgO/Fe sandwiches has been carried out by Butler *et al.* [11] using first principles calculations as well as Landauer formula. The sandwiches had Fe(100) and MgO(100) layers constituting the sandwich structure. The self-consistent calculations allowed for lattice and charge relaxations. Their results indicate that, (1) the symmetry of the Bloch states in the electrodes affect the tunneling conductance very strongly; (2) The rate of decay of states of different symmetry inside the barrier are quite different (d states decay more rapidly compared to s-like states); (3) There is the possibility of quantum interference inside the barrier, which can lead to oscillatory behavior as a function of $\mathbf{k}_\parallel$ and damped oscillatory behavior with barrier thickness; (4) Interfacial resonances may permit efficient tunneling and that the two spin channels may have completely different tunneling mechanisms associated with them. The latter is due to different types of (the two spin channel) states at the Fermi level. For example, a majority spin state with Δ_1 symmetry is said to be able to effectively couple from Fe into the barrier (MgO) and then out of the barrier into the second Fe electrode. In the minority channel, the conductance was found to be dominated by interface resonance states, in particular for thin MgO barriers.

Some of the insights obtained in the above study could be relevant to the general problem of tunneling. For example, although Fe has a relatively low density of majority d

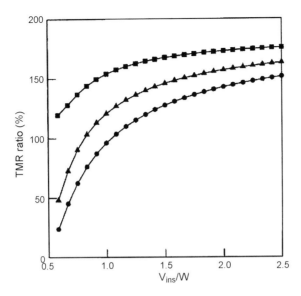

Figure 7.4. Dependence of the tunneling magnetoresistance on the height V_{ins} of an insulating barrier between the ferromagnetic electrodes for a barrier whose thickness is one (squares), three (triangles), and five (circles) atomic planes. A single orbital tight binding with band width W has been used here. Reproduced with the permission of Elsevier from Ref. [7].

states compared to the minority d states at the Fermi level, it was found that the s majority states are primarily responsible for tunneling in the majority channel. The reason for this is that the d states decay faster in the barrier when compared to the decay of the s states. In addition, in cases where a superconducting electrode could be used to determine conductance in a tunneling barrier, the reason may be simply due to the fact that such conductance in most (magnetic) transition metals is dominated by majority s states, rather that d states.

7.4. Devices with large TMR values

In order to realize high speed MRAM, it is highly desirable to have cells (MTJs) that have large TMR values. Following several theoretical predictions [11,12], MgO based MTJs with ferromagnetic electrodes have become a favorite set of devices to explore. Epitaxial Fe(001)/MgO(001)/Fe(001) MTJs provide an example of one of the basic architectures that has been fabricated in this regard in the experiments described in Refs. [13–15]. The measured magnetoresistance values in such MTJs have surpassed the values observed in conventional MTJs, reaching well over 100% at room temperature. All of these have been carried out with single crystal substrates, which can sometimes limit their practical use. Later, using polycrystalline FeCo or amorphous CoFeB electrodes with MgO as the insulating barrier, MR ratios of about 230% have been obtained [16]. However, more recently TMR values above 400% at room temperature and 804% at 5 K, have been reported with MgO barriers and Fe-Co-B based ferromagnetic electrodes [17,18]. Yuasa *et al.* [19] have also shown that magnetoresistance values exceeding 400% can be achieved using Fe-Co

Table 7.1   Maximum TMR ratio for MTJs with different reference and free layers annealed at optimum temperature T_a^{op} that yields the indicated room temperature TMR values

Reference layer	Free layer	T_a^{op} (°C)	TMR ratio %
$Co_{40}Fe_{40}B_{20}$	$Co_{40}Fe_{40}B_{20}$	400	355
$Co_{40}Fe_{40}B_{20}$	$Co_{50}Fe_{50}$	400	277
$Co_{40}Fe_{40}B_{20}$	$Co_{90}Fe_{10}$	350	131
$Co_{50}Fe_{50}$	$Co_{40}Fe_{40}B_{20}$	325	50
$Co_{50}Fe_{50}$	$Co_{50}Fe_{50}$	270	12
$Co_{90}Fe_{10}$	$Co_{90}Fe_{10}$	270	53
$Co_{90}Fe_{10}$	$Co_{40}Fe_{40}B_{20}$	300	75

Reproduced with the permission of Elsevier from Ref. [17].

electrodes and insulating MgO layers epitaxially matched at the interfaces. Some of these room temperature TMR values with Fe-Co alloyed electrodes are shown in Table 7.1 and Figure 7.5.

In most of these examples, parallel (magnetization) configuration appears to be the one with higher (differential) conductivity. However, there are exceptions that arise due to the electronic structure of the magnetic electrodes and bias voltage as illustrated in Figure 7.6. In half-metallic Huesler alloy Co_2MnSi (CMS) based MTJs such as the one shown in Figure 7.7, an asymmetric crossover behavior in conductivity appears at a negative bias voltage [20]. As evident from the figure, the half-metallic system has a gap in the minority density of states at the Fermi level. The dominant mechanism contributing to the increase in antiparallel conductivity is said to be arising due to tunneling from occupied (majority) states in $Co_{50}Fe_{50}$ into vacant minority states of CMS, as indicated by the solid line in Figure 7.7. Here, the electronic structure of CMS rather than that of $Co_{50}Fe_{50}$ is said to be directly responsible for the observed crossover behavior.

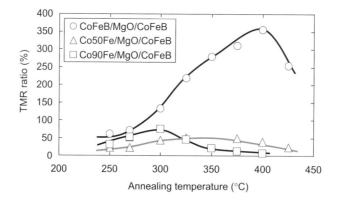

Figure 7.5.   Large TMR values obtained with MgO as the insulating barrier and their temperature dependence
Reproduced with the permission of Elsevier from Ref. [17].

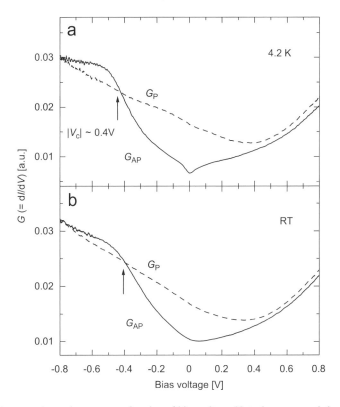

Figure 7.6. Asymmetric conductance as a function of bias voltage. Note the crossover behavior of G_P and G_{AP} at a negative bias voltage. Reproduced with the permission of Elsevier from Ref. [20].

7.5. Double barriers, vortex domain structures

Double barrier MTJs (see Figure 7.8) have also attracted attention due to their potential use in spin injection as well as detection [21]. Some such studies of double barrier devices have reported somewhat unexpected, low TMR values as evident from Figure 7.9 [22,23]. For example, observed TMR ratios of about 30%, compared to theoretically predicted values of 140% using $Co_{75}Fe_{25}$ were attributed to the occurrence of vortex domain structures and decreasing magnetization in the free layer [23]. The Julliére's formulas (for the single and double barrier MTJs) which were used to predict the TMR ratios were found to be valid only near 0 K and zero DC current (or bias voltage) when no vortex domain structures are present. Otherwise, dynamic magnetic domain structures appear to lower the TMR values according to Ref. [23].

Vortex structures have been observed using experimental techniques such as magnetic force microscopy (MFM) [24] and photoelectron emission microscopy PEEM [25]. A vortex structure has a chirality (clock-wise or counter clock-wise) as well as a polarity (up or down). MFM is able to detect the polarity of the core but not the chirality while PEEM can identify the chirality but not the polarity of the core. Vortex dynamics, for example under a magnetic field, can lead to various interesting phenomena not always beneficial

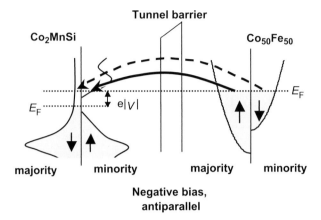

Figure 7.7.　Schematic diagram of the tunneling mechanism in a $Co_2MnSi/MgO/Co_{50}Fe_{50}$ MTJ at a negative bias voltage for the antiparallel magnetization configuration. Reproduced with the permission of Elsevier from Ref. [20].

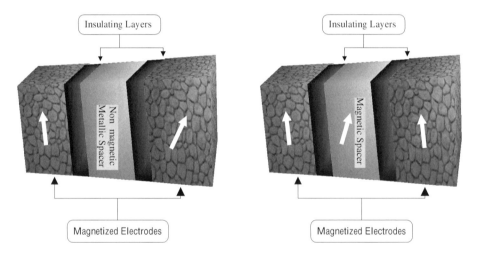

Figure 7.8.　Possible architectures of double barrier magnetic tunnel junctions.

to optimizing devices such as MTJs. Understanding and controlling their behavior would be crucial when designing efficient MTJs.

7.6.　Spin transfer torques in metallic multilayers

Normally, application of an external magnetic field can be expected to switch the magnetization of a magnetic layer, especially if it is somewhat "free" to orient itself. This external field may be produced by a current and is sometimes referred to as an Oersted field. However, when one considers the transport of electrons through a small metallic system (medium), it is interesting to pose the question "what kind of an effect will the

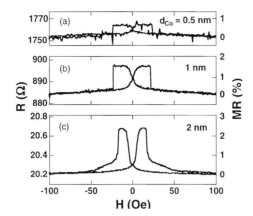

Figure 7.9. Magnetoresistance curves of $Ni_{80}Fe_{20}/Al_2O_3/Co/Al_2O_3/Co$ double barrier magnetic tunnel junction. Reproduced with the permission of Elsevier from Ref. [22].

current have on the medium itself?" When the medium consists of an interface of thin magnetic and nonmagnetic films, the situation becomes quite interesting. It turns out that the Oersted field is capable of polarizing the media.

In 1996, Slonczewski [26] and Berger [27] predicted such a fascinating effect in magnetic multilayers. The passage of electrons through a ferromagnet results in polarizing the electrons (i.e., transfer of (spin) angular momentum) which in turn will interact with neighboring non-magnetic media. Thus, sufficiently large currents would lead to spin precessions and large spin torques responsible for changes in magnetizations of the medium.

In order to make use of spin transfer in a multilayer system, it should be fabricated in such a way that in one set of layers, the magnetization remains unchanged while another set of layers responds to the spin current and undergoes changes in magnetization. The latter set is referred to as "free layers". We have encountered such situations in the discussion of "exchange-bias" in Chapter 4. A device which demonstrates such changes in magnetization is likely to exhibit changes in magnetoresistance as in GMR and TMR devices discussed in this volume and these could be utilized in various sensor and other applications.

Spin transfer effect, which has now been observed in many laboratories, is a well established phenomenon with a strong, underlying theoretical footing [28]. Various possible applications are now being considered such as Magnetic Random Access Memory devices. The magnetic state of the free layers, which can be controlled as mentioned above, can be used as the information that can be stored. A technique similar to that utilized in MRAM can be used to write data in a medium such as a magnetic hard disk. Here, a probe would write on to magnetic bits using current pulses while being scanned over the media. Another application could be in nanomechanical devices, such as nanomotors. A spin-polarized current can be used to transfer angular momentum to a nanomagnet and thereby induce rotational motion. In fact, several experiments have shown that a spin-polarized current passing through a nanomagnet can generate a dynamic response as a result of the spin torque of the conduction electrons [29–31]. The following discussion of spin transfer torques is based on the work of Stiles and Miltat [2].

Devices based on spin transfer torques share at least two characteristics; magnetoresistive readout of the magnetic state and small cross sectional area. The resistance of the device depends on its magnetic state, typically through the GMR effect, so that a measurement of the resistance yields information about the magnetic state. Frequently, the layers are fabricated so that there is a free layer, whose magnetization is free to respond to a current, and a layer with fixed (pinned) magnetization. Typical measurements are either the resistance V/I or differential resistance dV/dI as a function of current. The signature of a spin transfer effect is a change in the resistance of the device that is asymmetric in the current. With Oersted fields, this change is symmetric in the current.

Also, all of the devices have small cross sectional areas, since (a) fairly large currents are needed for angular momentum transfer and (b) the relative effect of Oersted fields decreases as the cross sectional area decreases. The simplest way to obtain a small cross sectional area is through a mechanical point contact. In these devices, a sharp tip is lightly pressed into a sample. The first observation of a spin transfer effect by Tsoi *et al.* (Ref. [32]) used such a mechanical point contact to a magnetic multilayer.

Peaks in the differential resistance (dV/dI vs. V) were observed only for one polarity of the current. This was regarded as evidence for spin transfer effect. The current at which the peak occurred increased linearly with increasing magnetic field. More controlled devices have been fabricated using nanolithography. In high magnetic fields, these continue to show the peak structure seen in mechanical contacts. However, in low fields there appears to be a switching between two (hysteretic) stable states. The magnetoresistivity of these states match those of the "parallel" and "antiparallel" magnetic alignments.

To understand spin transfer effects, one should first consider the role of interatomic exchange interactions (discussed in Chapter 2). A spin current naturally consists of moving spins (see Figure 7.10). Both spin and velocity are vectors and hence the spin current should be a tensor (an outer product of two vectors).

$$\mathbf{Q} = (\hbar/2)P\hat{\mathbf{s}} \otimes \mathbf{j}. \tag{7.7}$$

Here P stands for scalar polarization, which is dimensionless. From a quantum mechanical point of view, using particle wavefunctions $\psi_{i,\sigma}$, the spin and current densities are,

$$\mathbf{s}(\mathbf{r}) = \sum_{i\sigma\sigma'} \psi^*_{i\sigma}(\mathbf{r})\mathbf{S}_{\sigma,\sigma'}\psi^*_{i\sigma'}(\mathbf{r}) \tag{7.8}$$

and

$$\mathbf{Q}(\mathbf{r}) = \sum_{i\sigma\sigma'} \mathrm{Re}\big(\psi^*_{i\sigma}(\mathbf{r})\mathbf{S}_{\sigma,\sigma'} \otimes \hat{\mathbf{v}}\psi^*_{i\sigma'}(\mathbf{r})\big) \tag{7.9}$$

respectively, where $\hat{\mathbf{v}}$ and $\mathbf{S}_{\sigma,\sigma'}$ are velocity and spin operators.

To proceed further, let us look at the equations of motion for the number and spin densities. The number density satisfies the familiar continuity equation,

$$\frac{\partial n}{\partial t} = -\nabla \cdot \mathbf{j}, \tag{7.10}$$

which expresses the conservation of the number of particles, i.e., the rate of change of electron density in a given region is given by the net flow of electrons into that region.

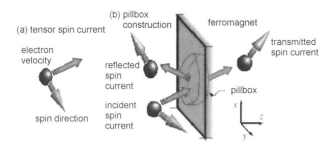

Figure 7.10.   (a) An electron moving in one direction with its spin in another direction illustrating a tensor spin current. (b) A pillbox around interface for computing the interfacial torque. Reproduced with the permission of M.D. Stiles [2].

The term on the right is arising from the noncommutativity of the kinetic energy term in the Hamiltonian with the number density.

The equation of motion for spin density, **s**, satisfies a similar equation with extra terms, arising from the noncommutativity of terms such as magnetocrystalline anisotropy in the Hamiltonian with the spin density.

$$\frac{\partial \mathbf{s}}{\partial t} = -\nabla \cdot \mathbf{Q} + \mathbf{n}_{ext} \tag{7.11}$$

where $\nabla \cdot \mathbf{Q} = \partial_k Q_{ik}$.

Equation (7.11) can be interpreted similar to the equation of continuity. The time rate of change of spin is given by the net flow of spin current into a given region and the contribution from the external torque density $\mathbf{n}_{ext}$, which tends to rotate the spins. Recall that in (classical) Newtonian mechanics, the rate of change of angular momentum about an axis is given by the sum of the external torques with respect to that axis. There are two contributions to the spin current. First, the mediator of the exchange interaction, which is carried by all of the electrons contributing to the magnetization. If the magnetization direction is non-uniform in a ferromagnet (FM), the (left and right) spin currents carried by the eigenstates do not cancel. The gradient of this current gives rise to a spin torque that tends to rotate the non-uniform magnetization as in $\mathbf{n}_{ex} = -\nabla \cdot \mathbf{Q}_{ex}$.

Second, the spin current which is generated by an imbalance in the populations of the spin states at or near the Fermi energy. This is strictly a nonequilibrium spin current due to the imbalance of left and right moving spins and will be the focus of the rest of this section. If the spin density is converted to magnetization, the above equation becomes the Landau–Lifshitz–Gilbert (LLG) equation without damping. There is no damping here since the equation was written for a single particle Hamiltonian. If many electron effects are included, then damping terms will appear in this equation. Experimentally, there have been numerous efforts to verify the applicability of the LLG equation and to monitor damping parameters. For example, recently, current induced reversals of magnetization in permalloy/Cu/permalloy nanopillars have been reported at temperatures $T = 4.2$ K to 160 K [33] and the damping has been found to be strongly temperature dependent.

The spin dependent reflection (or transmission) is crucial to understanding the behavior of spin currents at interfaces. The Fermi surfaces and hence scattering properties are different for up and down electrons. Consequently, the reflection and transmission amplitudes of electrons arriving from the nonmagnet (NM) and scattering at the (NM/FM) interface will depend on whether their spins are parallel or antiparallel to the magnetization in the ferromagnet.

The torque is determined by the reflection or transmission of electrons at an arbitrary angle to the magnetization. The behavior of such electrons can be described as follows. The state of an electron with a spin at a polar angle θ and an azimuthal angle ϕ is a superposition of the spin up and spin down states, which can be written as $e^{ikz}|\theta, \phi\rangle$ where,

$$|\theta, \phi\rangle = \cos(\theta/2)e^{-i\phi/2}|\uparrow\rangle + \sin(\theta/2)e^{i\phi/2}|\downarrow\rangle. \tag{7.12}$$

Now the reflection coefficients, $R_\uparrow$ and $R_\downarrow$, which will in general be unequal at a FM/NM interface, will yield a different state compared to the incoming one shown above; i.e.,

$$e^{-ikz}\left\{R_\uparrow \cos(\theta/2)e^{-i\phi/2}|\uparrow\rangle + R_\downarrow \sin(\theta/2)e^{i\phi/2}|\downarrow\rangle\right\}. \tag{7.13}$$

Here, note that the reflected electron is rotated with respect to the incoming electron. The new polar angle θ' is determined by $\tan(\theta'/2) = |R_\downarrow/R_\uparrow| \tan(\theta/2)$ and the new azimuthal angle is determined by $\phi' = \phi + \text{Im}\{\ln(R_\uparrow^* R_\downarrow)\}$.

The transmitted electrons will see an added complication. Inside the ferromagnet, the spin up and down electrons have different **k** vectors, $\mathbf{k}_\uparrow$ and $\mathbf{k}_\downarrow$. One consequence of this is a phase factor $e^{i(k_\uparrow - k_\downarrow)z}$ which changes as the electron moves further inside. This results in a change of the azimuthal angle, which is a manifestation of spin precession. Figure 7.11 shows a cartoon depicting this situation.

To understand the effects of non-collinear spin scattering, consider a simple model proposed by Waintal *et al.* [34], in which the reflection probability for minority spins is one and for majority spins is zero. Consider an electron having a spin perpendicular (along $\hat{\mathbf{z}}$) to the moment of the ferromagnet (along $\hat{\mathbf{x}}$) scattering off the interface shown in Figure 7.11. This incoming electron is in a state that is a superposition of majority and minority spin states. Once scattered, all of the majority amplitude is transmitted into the ferromagnet while all of the minority amplitude is reflected back into the non-magnet. The spin currents in this situation are given by,

$$\mathbf{Q_{in}} = (\hbar/2)\hat{\mathbf{z}} \otimes v\hat{\mathbf{z}},$$
$$\mathbf{Q_{refl}} = (\hbar/4)(-\hat{\mathbf{x}}) \otimes (-v\hat{\mathbf{z}}),$$
$$\mathbf{Q_{trans}} = (\hbar/4)\hat{\mathbf{x}} \otimes v\hat{\mathbf{z}}. \tag{7.14}$$

The total change in spin flux, $\mathbf{N}_c$ in a pillbox of Figure 7.10, is given by

$$(\mathbf{Q_{in}} - \mathbf{Q_{trans}} + \mathbf{Q_{refl}}) \cdot A\hat{\mathbf{z}} = Av\hbar\hat{\mathbf{z}}/2. \tag{7.15}$$

This example illustrates two key features of the scattering process. The first, which is valid in general, is that the spin current along the direction of the magnetization is conserved. The second point (which is only approximately true in general) is that the reflected and transmitted currents have no transverse components. When combined, these two results show that the spin transfer torque is approximately given by the absorption of the transverse component of the incident spin current. The mechanism for this transfer of angular momentum is the exchange interaction experienced by the electrons in the ferromagnet, which exerts a torque on the spin current. This, in turn, exerts a reaction torque

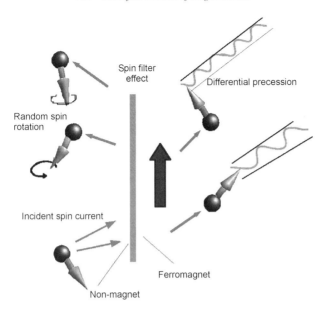

Figure 7.11. Mechanisms contributing to absorption of incident transverse spin current. Electrons incident from the non-magnet (lower left) are distributed over a distribution of states (two of which are shown). All these electrons are in the same spin state, which is transverse to the ferromagnetic spin density (shown by the large arrow). The reflected electron spins have predominantly minority character and their transverse components are distributed over many directions because of the variation over the Fermi surface of the phases of the reflection amplitudes. The transmitted electron spins precess as they go into the ferromagnet since the wave vectors for the majority and minority components are different. Electrons with different initial conditions precess at different rates, leading to classical dephasing (differential precession). Reproduced with the permission from M.D. Stiles [2].

on the ferromagnet. Although the above model is not valid in general for realistic cases, its conclusions are pertinent to transition metal multilayers such as Co/Cu [28,35]. The spin filter effect here essentially separates the majority and minority spin currents, no longer interfering with each other, which leads to the vanishing of transverse components in the outgoing current. As depicted in Figure 7.11, when the reflected spins carry a transverse component, it would be rotated with respect to the incident one (as shown in Equation (7.13)). It turns out that, when summed over the Fermi surface, the reflected transverse component largely cancels out (which is due to random spin rotation, an example of classical dephasing). The above discussion is only a simple introduction to spin scattering at a ferromagnetic interface and the interested reader is referred to the growing list of references on this topic, some of which can be found at the end of this chapter.

7.7. Ultra-fast reversal of magnetization

In order to establish its place and compete effectively in the solid-state memory market, it is necessary for the future MRAM devices to operate in the GHz range (i.e., sub-nanosecond switching times) with cell sizes under 100 nm. Spin torque based "pseudo spin valves", which consist of a CPP (current perpendicular to plane) GMR device, patterned into pillar geometry (Figure 7.12(A)), have been reported to operate in the sub-nanosecond regime [36]. In fact, in this experiment, spin transfer torques have been shown to trigger reversals of magnetization in the 100 picosecond (ps) regime. Starting from either the parallel (P) or antiparallel (AP) configurations, a pulse either switches completely or does not switch the magnetization. Between 50 and 300 K, after-pulse resistances were R_{AP} or R_P with nothing in between. An important observation here is that the pulse-induced reversal is probabilistic when the pulses are short but strong enough to trigger deterministic switching. Fast and reliable switching was achieved at moderate pulse amplitude while the fastest reversal turned out be the most cost-effective. They also claim to have demonstrated speed scalability; i.e., increasing the switching speed reduces the energy cost of reversal.

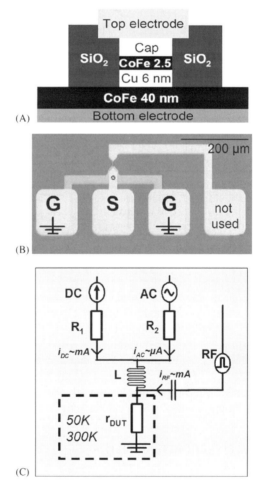

Figure 7.12.   (A) Cross section of the CPP GMR device under test for fast switching. Thicknesses are in nm. Lateral size is 85×150 nm^2. (B) Optical micrograph of the high bandwidth accesses to the device (central dot). (C) Sketch of the measurement set up. Reproduced with the permission of Elsevier from Ref. [36].

Another way to achieve ultra-fast switching of magnetization is by applying specifically shaped magnetic field pulses that match the intrinsic properties of the magnetic elements. In principle, the fundamental limit of the switching speed is determined by (half of) the precession frequency. However, due to what is referred to as "ringing" or underdamping, the relaxation of the system takes much longer than a few precessional periods. In Ref. [37], it is reported that switching times of about 200 ps can be achieved using magnetic pulses as described above. Similar studies, using spin-valve structures and hard-axis pulses that achieve ultra-fast switching, have also been reported recently [38,39]. Further extensive studies, that examine issues such as those due to temperature changes and damping [40], are needed before such spin-valves would become practical, but this area of research looks very promising.

7.8. Transistors based on spin orientation

Spin transistors use two ferromagnetic layers, one as the "spin injector" and the other as the "spin analyzer". These have unique output characteristics which depend on the relative orientation of the two magnetizations of the ferromagnetic layers as well as the bias conditions. As an example, Ref. [41] describes a magnetic tunneling transistor (MTT) which operates at room temperature. This MTT consists of a ferromagnetic emitter (CoFe), a tunnel barrier (Al_2O_3), an ultra-thin ferromagnetic base layer (CoFe) and a collector in the form of a GaAs(111) substrate. The spins injected by the emitter are filtered by the base (FM) layer. Hence the collector current is quite sensitive to the spin orientation of the ultra-thin base layer. Output (collector) currents exceeding 1 μA and current changes up to 64% (due to changes in magnetization) have been observed in this device. There have been other attempts or proposals using MTT ideas such as in the metal-oxide-field-effect-transistor using half metallic contacts for source and drain [42]. The motivation behind such attempts is that these can be used as cells for non-volatile binary data and for non-volatile, reconfigurable logic based on spin transistor gates.

7.9. Summary

The topics discussed in this chapter have to do with manipulating and detecting spin-polarized currents. These ideas have their roots going back to the 1930s (Ref. [43]), when it was realized that spin polarized transport consists of two mostly independent channels pertaining to majority and minority spins. During the past two decades, since the discovery of the GMR effect, this field of spintronics has blossomed into a very robust and active area of research with both fundamental and practical implications. Spintronic devices such as spin valves have had a significant impact on magnetic recording hard disk drives leading to increases in capacity by at least a couple of orders of magnitude since they were introduced by IBM [44]. The MTJs which were engineered subsequently have become quite competitive in providing sufficiently fast, non-volatile memory. The first generation of such MRAM devices, having a capacity up to 4 Mbits, are commercially available [45]. With decreasing cell size, switching the magnetizations from a distance could be problematic. However, spin transfer torques may provide a key solution due to its more local switching mechanism. Spin transfer devices that use MgO-based MTJs look quite promising in this regard [46]. All these developments are very likely to lead to revolutionary advances in MRAM such as instant boot-up computers. Most of the spintronic devices discussed in this volume are either metallic multilayers or multilayer sandwich structures with insulating films (i.e., tunneling devices). At nanoscale thicknesses, some of the bulk-like properties are found to be modified and understanding such changes, effects due to applied fields, temperature as well as effects at the interfaces have been essential to the rapid development of this field.

References

[1] Press release, NEC Corporation (Feb. 7, 2007).
[2] M. Stiles, J. Miltat, Private communication.
[3] G.H. Koh, et al., J. Magn. Magn. Mater. **272–276**, 1941 (2004).
[4] R. Meservey, P.M. Tedrow, P. Fulde, Phys. Rev. Lett. **25**, 1270 (1970);
    R. Meservey, P.M. Tedrow, Phys. Rev. Lett. **26**, 192 (1971).
[5] M. Julliere, Phys. Lett. A **54**, 225 (1975).
[6] J.S. Moodera, L.R. Kinder, T.M. Wong, R. Meservey, Phys. Rev. Lett. **74**, 3273 (1995).
[7] J.S. Moodera, G. Mathon, J. Magn. Magn. Mater. **200**, 248 (1999).
[8] J. Slonczewski, Phys. Rev. B **39**, 6995 (1989).
[9] R. Landauer, IBM J. Res. Dev. **32**, 306 (1988).
[10] See for example, G.D. Mahan, Many-Particle Physics (Plenum, New York, 1981).
[11] W.H. Butler, X.-G. Zhang, T.C. Schulthess, Phys. Rev. B **63**, 054416 (2001).
[12] J. Mathon, A. Umerski, Phys. Rev. B **63**, 220403R (2001).
[13] J. Faure-Vincent, et al., Appl. Phys. Lett. **82**, 4507 (2003).
[14] S. Yuasa, A. Fukushima, T. Nagahama, K. Ando, Y. Suzuki, Jap. J. Appl. Phys. Part 2 **43–4B**, L588 (2004).
[15] S. Yuasa, T. Nagahama, A. Fukushima, Y. Suzuki, K. Ando, Nat. Mater. **3**, 868 (2004).
[16] D.D. Djayaprawira, K. Tsunekawa, M. Nagai, H. Maehara, S. Yamagata, N. Watanabe, S. Yuasa, Y. Suzuki, K. Ando, Appl. Phys. Lett. **86**, 092502 (2005).
[17] S. Ikeda, J. Hayakawa, Y.M. Lee, F. Matsukara, H. Ohno, J. Magn. Magn. Mater. **310**, 1937 (2007).
[18] J. Hayakawa, S. Ikeda, Y.M. Lee, F. Matsukara, H. Ohno, Appl. Phys. Lett. **89**, 232510 (2006).
[19] S. Yuasa, et al., Appl. Phys. Lett. **89**, 042505 (2006).
[20] T. Marukame, et al., J. Magn. Magn. Mater. **310**, 1946 (2007).
[21] S. Stein, R. Schmitz, H. Kohlstedt, Solid State Commun. **117**, 599 (2001).
[22] H. Kubota, T. Watabe, T. Miyazaki, J. Magn. Magn. Mater. **198–199**, 173 (1999).
[23] X.F. Han, et al., J. Magn. Magn. Mater. **282**, 225 (2004).
[24] T. Shinjo, et al., Science **289**, 930 (2000).
[25] S.-B. Choe, et al., Science **304**, 420 (2004).
[26] J. Slonczewski, J. Magn. Magn. Mater. **159**, L1 (1996);
    J. Slonczewski, J. Magn. Magn. Mater. **195**, L261 (1999).
[27] L. Berger, Phys. Rev. B **54**, 9353 (1996);
    L. Berger, J. Appl. Phys. **90**, 4632 (2001).

[28] M.D. Stiles, A. Zangwill, Phys. Rev. B **66**, 014407 (2002).

[29] S.I. Kiselev, J.C. Sankey, I.N. Krivorotov, N.C. Emley, R.J. Schoelkopf, R.A. Buhrman, D.C. Ralph, Nature (London) **425**, 380 (2003).

[30] S. Urazhdin, et al., Phys. Rev. Lett. **91**, 146803 (2003).

[31] J.A. Katine, et al., Phys. Rev. Lett. **84**, 3149 (2000).

[32] M. Tsoi, A.G.M. Hansen, J. Bass, W.C. Chiang, M. Seck, V. Tsoi, P. Wyder, Phys. Rev. Lett. **80**, 4281 (1998).

[33] N.C. Emley, et al., Phys. Rev. Lett. **96**, 247204 (2006).

[34] X. Waintal, E.B. Myers, P.W. Browser, D.C. Ralph, Phys. Rev. B **62**, 12317 (2001).

[35] K. Xia, et al., Phys. Rev. B **65**, 220401 (2002).

[36] T. Devolder, A. Tulapurkar, K. Yagami, P. Crozat, C. Chappert, A. Fukushima, Y. Suzuki, J. Magn. Magn. Mater. **286**, 77 (2005).

[37] Th. Gerrits, H.A.M. van den Berg, J. Hohlfeld, L. Bär, Th. Rasing, Nature (London) **418**, 509 (2002).

[38] S. Kaka, S.E. Russek, Appl. Phys. Lett. **86**, 2958 (2002).

[39] W.K. Hiebert, G.E. Ballentine, M.R. Freeman, Phys. Rev. B **65**, R140404 (2002).

[40] G.D. Fuchs, J.C. Sankey, V.S. Pribiag, L. Qian, P.M. Braganca, A.G.F. Garcia, E.M. Ryan, Z. Li, O. Ozatay, D.C. Ralph, R.A. Buhrman, Appl. Phys. Lett. **91**, 062507 (2007).

[41] S. van Dijken, X. Jiang, S.S.P. Parkin, Appl. Phys. Lett. **80**, 3364 (2002).

[42] S. Sugahara, M. Tanaka, Appl. Phys. Lett. **84**, 2307 (2004).

[43] N. Mott, Proc. Roy. Soc. **156**, 368 (1936).

[44] S.S.P. Parkin, X. Jian, C. Kaiser, A. Panchula, K. Roche, M. Samant, Proceedings of the IEEE **91**, 661 (2003).

[45] Press release, Freescale semiconductor (July 10, 2006).

[46] C.L. Chien, F.Q. Zhu, J. Zhu, Physics Today, June 2007.

Chapter 8

Confined Electronic States in Metallic Multilayers

Ultrathin metal films grown epitaxially on metal substrates and related multilayered sandwich structures are prime candidates in the search for confined states. In bulk metallic systems, electrons have the freedom to move throughout the 3-dimensional space. When the size of a given metallic system is reduced to nanometer scale along at least one direction, it is natural to expect some electron confinement along that direction. These electronic states may be described as resonances arising from interface/surface reflectivity. As we have already seen and discussed in this volume, such confined states are likely to show effects not seen in the bulk structures and produce size dependent responses to external and other probes. These may be viewed as some sort of a "resurrection" of otherwise "dead" materials science problems. In a sandwich structure, where magnetic layers are separated by a nonmagnetic spacer, we can expect to see confined electronic states, sometimes referred to as quantum well (QW) states, when the spacer layer thickness satisfies appropriate boundary conditions. Since the early nineties, similar metallic QW states have been detected and carefully examined (Refs. [1–8]). Although these confined states were not an essential part of the original RKKY explanations of the GMR effect, some of the observed oscillatory phenomena tied to exchange coupling can be easily linked to such states, as discussed in Chapter 1.

There is a simple analog of confined states in mechanical systems, namely standing waves. When the wavelength of the vibrations satisfies an appropriate condition tied to the confining length, standing (or stationary) waves can be observed in a vibrating string. It can now be argued that depending on the repeat length (period) of the multilayer system, one can expect oscillatory variations in electronic properties. Fermi surfaces of metallic elements such as Cu, exhibit various shapes (neck, belly etc.) along various directions in **k**-space. A natural question that arises is whether there are other physical observables (for example, work function of multilayers) that exhibit clear oscillations with respect to the period of the multilayer system.

In this chapter, we will further discuss experimental evidence for the existence of confined states in multilayer systems and examine how far can we take simple theories of spin-dependent transport, provided **k**-resolution is available. Using high quality single crystal samples, it may be possible to obtain (and possibly tune) **k**-dependent transport. It is instructive to review some of the past work in this field, in order to understand the historical progression. We begin with an ARPES study (see Chapter 5 for a discussion of ARPES), a probe of local density of states in **k**-space and related physical phenomena due to Qiu *et al.* [9] using the advanced light source (ALS), a third generation synchrotron light source at Lawrence Berkeley National Laboratory, with wedge shaped samples. In ARPES, when probing valence states, an incident photon is used to kick electrons out from occupied valence levels, and their final energies are measured. The component of electron momentum parallel to the surface is conserved (assuming that the momentum of the incident photon is negligible) to within a reciprocal lattice vector which allows a measure of the binding energies as a function of the parallel momentum of the electron (see the chapter on experimental probes in this volume for more details).

8.1.   Wedge shaped samples

As discussed previously in Chapter 1, there are resonances and other features of multilayer systems which depend on the thickness of the spacer layers. It is clearly desirable to make thickness dependent measurements under *the same experimental conditions*. The wedge shapes used in Ref. [9] (Figure 8.1) provide a clever and convenient way of probing different thicknesses under identical conditions, provided one is able to target a narrow region of a given sample where the thickness of the wedge does not change substantially; i.e., the diameter of the area illuminated by synchrotron radiation should be small so that the thickness variations of the wedge can be ignored. The size of the photon beam used is 50–100 µm with a high enough photon flux for the ARPES work. Thus for a wedge with a slope of 5 ML/mm, a scan of a 50 µm photon beam will provide a $50 \times 10^{-3} \times 5 = 0.25$ ML thickness resolution.

In addition to the high photon flux and the narrow beam, X-ray synchrotron radiation provides access to the core levels. This enables X-ray magnetic linear dichromism (XMLD) studies to be carried out, which can be used to determine the direction of magnetization of a layer. The asymmetry in the X-ray photoemission intensity I in various directions is a manifestation of the changes in magnetization induced by the spin-orbit interaction along those directions. Hence the XMLD asymmetry, defined as $[I(+M) - I(-M)]/[I(+M) + I(-M)]$ with M being the total magnetization, is sensitive to the direction of magnetization in the sample even though the X-rays probe the core levels. This measurement yields information about the sign of magnetic interlayer coupling near the surface region. To determine the strength of this coupling, another technique (such as SMOKE) has to be used.

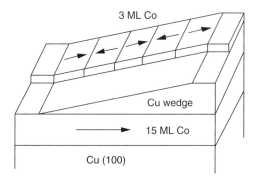

Figure 8.1.   Wedge-shaped sample used in the photoemission study of Ref. [9]. The overlayer of Co was deposited over half the sample. Note how the magnetization alternates in this overlayer with respect to the Co substrate; this is due to the exchange coupling that depends on the thickness of the Cu spacer. The XMLD measurements of the Co covered half can be correlated with the belly and neck (Figure 8.2) periodicities obtained in photoemission from the uncovered half. Reproduced with the permission from the Institute of Physics.

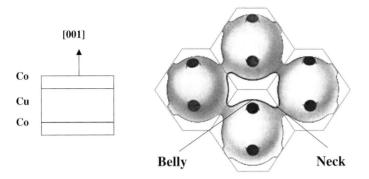

Figure 8.2.   Schematic drawing of the Co/Cu/Co(001) sandwich and the Fermi surface showing the belly and neck region [9]. Reproduced with the permission from IOP, the Institute of Physics.

Magneto Optic Kerr Effect (MOKE) makes use of Kerr rotation of the plane of polarization of light due to the magnetic anisotropy of a given medium. This has been discussed in some detail in Chapter 4; the essential point is that when light interacts with a magnetic medium, its plane of polarization can change (rotate) and this rotation carries information about the direction of magnetization of the medium.

Single quantum wells may be used to obtain a qualitative understanding of resonances and other properties, such as confinement energies of QW states, as shown in Chapter 1. Also, room temperature measurements are expected to yield useful results since the QW states can be observed at room temperature, which implies that the thermal energies are smaller than the quantized energy splittings under observation. However, it turns out that the nanometer size has to be controlled to a high precision when the multilayers are grown. Free electron models are widely used in order to discuss multilayer QWs due to their simplicity (and separability). However, there is no a priori reason for their use in noble and transition metals, where the Fermi surface consists of d electrons. Some of their limitations will be discussed later in this chapter. Assuming a free electron band structure in a single QW having width d and infinite walls, one can easily write down the energy of a confined electron (state) as follows;

$$E_n = \frac{\hbar^2 k^2_{\parallel}}{2m} + \frac{\hbar^2 k_z^2}{2m} = \frac{\hbar^2 k^2_{\parallel}}{2m} + \frac{\hbar^2 \pi^2 n^2}{2md^2}. \tag{8.1}$$

This equation illustrates the wavevector $k_z = \frac{n\pi}{d}$ associated with the direction of confinement z, with n being the number of half-wavelengths that can be fitted into a (QW of) width d. The energy of such a state decreases with increasing width. Here, the boundary conditions are such that the electron wave function vanishes at the two boundaries. However, for a multilayer system QW, the boundary conditions will be different.

8.2.   Phase accumulation model

In order to examine the single QW state energies and their dependence on the film thickness, a simple approach, labeled phase accumulation model (PAM) [10], has been employed during the past decade. Originally used to describe QW states in Ag/Fe(001) [11]

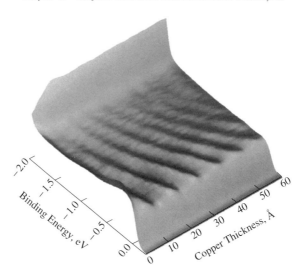

Figure 8.3.   Photoemisson spectra taken along the surface normal corresponding to the belly direction of the Cu Fermi surface. Oscillations in intensity as a function of the Cu thickness and electron energy show the formation of QW states in the Cu layer. Reproduced with the permission of the Institute of Physics [9].

and subsequently applied to multilayer sandwiches [5], this is one of the basic quantum well models one encounters in elementary quantum mechanics. The statement that the period with which the QW states cross the Fermi level is solely determined by the spacer layer thickness is a natural consequence of this model. It chooses a fixed $\mathbf{k}_\parallel$ in the 2-dimensional Brillouin zone and focuses on the projected band structure as a function of k_z, the perpendicular component of the Bloch vector. The phase shift of an electron upon reflection from each boundary is taken into account together with that due to the distance traveled. For a resonance to occur, these wave functions must interfere constructively. Hence, the quantization condition for an electron in a simple (rectangular box) potential well of width d can be written as

$$2k_z d_{Cu} + \phi_C + \phi_B = 2\pi n, \tag{8.2}$$

where ϕ_C and ϕ_B are the phase changes (gains) of the one electron wave function at the boundaries of the well, n (again) is the number of half-wavelengths and k_z is the wavevector (component along the z direction) associated with the electronic state. Depending on the region of the Fermi surface that is relevant to the QW states, the energy dispersion can be quite different for different states. Now for energies of interest and for fixed n, increasing the width of the well lowers the energy relevant to the k values in the above equation. However, the experimental observations for the system under study (Cu spacers) show that the energy associated with QW states increase with increasing width and the oscillation periodicity (5.88 ML). The latter matches with the long-period magnetic interlayer coupling in the Co/Cu/Co(001) system, when the analytic form $\pi/|k_{BZ} - k_F|$ is used for k_z.

Using this form, Equation (8.2) can now be written as

$$2k^e d_{Cu} - \phi_C - \phi_B = 2\pi \nu, \tag{8.3}$$

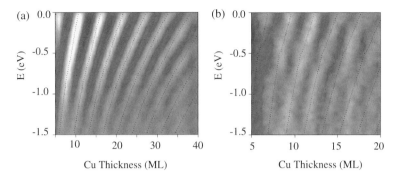

Figure 8.4. Photoemisson intensity along the (a) belly directions, (b) neck direction of the Cu Fermi surface showing different periodicities with Cu thickness. Dotted curves are results from PAM (see text). Compare with Figures 1.7 and 1.8 as well. Reproduced with the permission of the Institute of Physics [9].

where $k^e = k_{BZ} - k$ decreases with increasing energy, and hence, the QW also increases with increasing width of the QW. Here $k^e = \frac{\pi}{(k_{BZ}-k_F)}$, $k_{BZ} = \pi/a$ and $v = m - n$. Due to this formulation, we now see an increase of the QW energy with increasing d_{Cu} and an oscillation period of $\pi/(k_{BZ} - k_F)$ at the Fermi level, as observed experimentally.

The authors of Ref. [9] examined why the latter equation was preferred and whether there was any physical significance attached to it. Although they do not support the idea that an envelope function is essential to the formation of the QW states, they describe the importance of k_e to an energy proximity effect. When the thickness of the Cu layers is increased from m to $m + 1$ layers, the new QW state with $n + 1$ half-wavelengths in the well is close in energy to the old state with n half-wavelengths. With the blurring due to imperfect thickness control, it is concluded that the constant $m - n$, rather than the constant n, will merge into continuous curves seen in Figures 8.3 and 8.4.

A similar study [12] was conducted for several different ferromagnetic (FM) elements keeping Cu as the spacer, namely symmetric fcc (001) sandwiches M/Cu/M where M = Fe, Co, and Ni. This study referred to the fact that both experimental and theoretical evidence [1,2,13,14] showed that the resonances were mainly confined to the spacer layer, even with relatively modest reflectivities at the boundaries. Using bulk Cu band structure, band energies $E(k_z)$ were obtained and inverted to yield $k_z(E)$, which would then give rise to an energy dependent phase shift. This model was then used to show how successive QW states cross the Fermi level with a thickness period that is determined by the nonmagnetic Cu spacer and not the ferromagnetic layers. The calculated periods of 5.7 ML at the belly ($\mathbf{k}_{\parallel} = 0.0$ Å^{-1}) and 2.6 ML at the neck ($\mathbf{k}_{\parallel} = 0.98$ Å^{-1}) (see Figure 8.2) are in fairly good agreement with the experimentally observed periods of oscillatory magnetic coupling as well as the periods QW states that cross the Fermi energy, measured with photoemission and inverse photoemission [2,4,13,14]. In addition, the model showed how the projected band gaps of the FM layers affect the electronic states of the nonmagnetic Cu spacer. Whenever a band gap is encountered in the FM layers, the energy dependence of the phase shifts upon reflection at the Cu/FM interface is strongly modified. The number of Cu spacer layers necessary for a QW state to cross the Fermi level is larger when a FM gap is located below the Fermi energy than when it is located above. At the belly of the Cu Fermi surface, the gaps in Co and Ni are below the Fermi level and hence no thickness shifts of the Fermi level crossings or of the long period coupling are observed. However, at the neck of the Cu Fermi surface, the gap of Fe is above the Fermi energy, for Co it spans E_F and for Ni it is below E_F. As a consequence, the model predicts that the short period of oscillation will occur at successively larger film thicknesses as the FM material is changed from Fe to Co to Ni.

8.3.   Interfacial roughness

One can understand, at least qualitatively, why ideal interfaces would be amenable to a model such as PAM based on stationary wave phenomena and phase accumulation. For rough interfaces, do the changes in phase at the interface affect the oscillatory behavior of the multilayer system? When interfacial roughness is taken into account, it is not obvious why such a model should work. This problem was experimentally examined in Ref. [9], where several experiments were conducted with varying degrees of interfacial roughness. In one experiment, the QW states in Cu films grown on Co and Ni (with different degrees of roughness) were compared. Here the same Cu wedge was grown on Co and Ni substrates in order to obtain a direct comparison; i.e., the substrate is one half Co and one half Ni. On this substrate, a Cu wedge was grown on top of both Co and Ni. (It was verified that Cu/Ni and Cu/Co have different degrees of roughness.) Photoemission measurements were done on the Cu wedge so that the QW state at the belly of the Fermi surface could be detected. The photoemission at the Fermi level was monitored for both Cu/Ni and Cu/Co as a function of layer thickness and the data indicated quite similar images for the two cases; i.e., QW states are observed at the same Cu layer thicknesses for both, supporting the notion that there is no phase shift from one to the other. In the second experiment, Co/Cu/Co(001) and Co/Cu/Ni(001) sandwiches were annealed to introduce interfacial mixing and MXLD signals were monitored. The long period of oscillations for both systems remained the same indicating that the different degrees of roughness did not affect the phase accumulation relevant for this period, which shows that there is no phase shift between the two systems. However, the short period present in Co/Cu/Co(001) was not present in Co/Cu/Ni(001), indicating a rougher interface in the latter. With varying annealing conditions to induce more roughness (such as higher temperatures, longer annealing times), it was possible to observe the vanishing of even the long period oscillations. An important conclusion from this study is that the coupling strength appears to depend on the interfacial roughness while the phase changes do not show such behavior.

8.4.   Envelope functions of the QW state

By focusing on double-wedge samples, the authors of Ref. [9] were able to directly monitor the envelope modulation of the QW state. This is said to be similar to examining a standing wave present in a string and probing at various nodal and anti-nodal type positions to detect its response The conclusion was that the envelope function simply modifies the QW state, and is secondary to its existence. The QW state is mainly determined by the quantization condition which essentially depends on the thickness of the spacer layers. A double-wedge sample with the wedges tapering off at right angles to one another and a thin probing layer (Ni in this case) in between provides a way to "see" the QW state

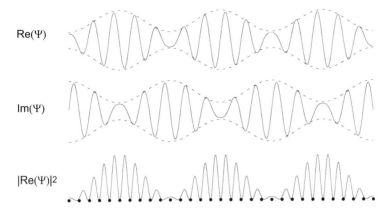

Figure 8.5. Wave functions in a corrugated box. On moving away from the BZ boundary the QW wave function gets modulated by an envelope function (dashed curves) characterized by a wavevector $2k^e$ (see text). The wave function is enhanced in p-like regions and depressed in s-like regions due to these modulations. Reproduced with the permission of the Institute of Physics [9].

(probabilities) at various points in space similar to a STM detecting the surface morphology using a tip. For example, this arrangement allows one to keep the thickness of the spacer constant and probe from one end of the QW to the other or keep the probing layer exactly in the middle of the well and monitor its effect on the QW resonances as the thickness of the spacer is varied. This technique enabled Qiu *et al.* [9] to observe modulations introduced by the envelope functions and understand their effects in terms of a "particle in a corrugated box" (see Figure 8.5).

8.5. Multiple quantum wells

Recently, Wu *et al.* [15] have further demonstrated that resonant coupling QW states can be realized by manipulating the energy levels of the barrier layers. First, they studied the double well Cu/Ni(2 ML)/Cu where the QW states near the Fermi level were shown to undergo a splitting due to the overlap of the wave functions from the two (Cu) QWs. However, each pair of the split energies merge together to form a single, degenerate state at energies below -0.5 eV (see Figures 8.6 and 8.7). This shows that the 2 ML thick Ni film is thick enough to isolate the two QWs and hence make the overlap of wave functions negligible at low energies.

For the triple QW system, Cu/Ni(1 ML)/Cu(14 ML)/Ni(1 ML)/Cu, the energy spectra show a different behavior when compared with the single and double well spectra [15]. Each previously degenerate state splits into two states when the energy E (measured with respect to the Fermi level E_F) is such that $-1.3 < E < -0.5$ eV (see Figure 8.8). The origin of the latter splitting must be connected with the center (14 ML) Cu film but is not due to a state crossing the Fermi level. It must be due to a state present in the center Cu layers that mediates the interaction between the two outer Cu films. This state can be called a resonant interaction channel. When no such state is present (at some other energy), the outer Cu layers do not see each other and no resonance would be observed. Indeed, as seen in Figure 8.8, when $-0.5 < E$ or $E < -1.3$, there are no middle layer QW

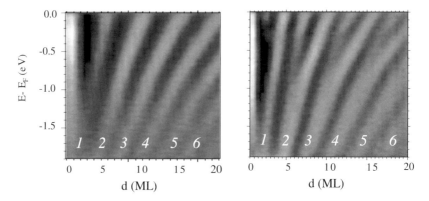

Figure 8.6.   Photoemission spectra from a symmetrical double well (right) as compared with the single well (left). Each degenerate QW state of the double well at low energy splits into two states at high energy with even and odd parities. Reproduced with kind permission of the Institute of Physics from Ref. [9].

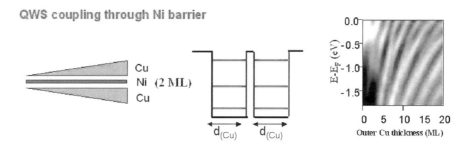

Figure 8.7.   A symmetrical double well structure and related photoemission spectra. Each degenerate QW state of the double well at low energy splits into two states at high energy with even and odd parities. Reproduced with kind permission from Y.Z. Wu. Also published in Ref. [15].

states present, which closes the above mentioned interaction channel. This observation suggests the possibility that by changing the middle layer thickness, the interaction energy may be tuned to a desired value, opening up a new way to manipulate QW structures.

8.6.   Non-free-electron-like behavior

An important issue concerning the QW states in real multilayer films has to do with their non-free-electron behavior and thickness dependent evolution. Again for the Cu/Co/Cu(001) system, systematic $k_\parallel$-dependent measurements of the Fermi surface, as well as the band structure of the QW states, show non-trivial effects away from normal emission [16]. The electronic structure of the Co barrier, such as various symmetry gaps, has been found to be responsible for some of these dramatic changes. For example, for an incoming electron with symmetry Δ_1 in Cu encounters no bands of similar symmetry in Co in the relevant energy range, then it may have to tunnel through an energy barrier in Co to travel through the Cu/Co/Cu(001) system. In addition, instead of states along ΓX, if states along XW are monitored, only even numbered intensity peaks in photoemission

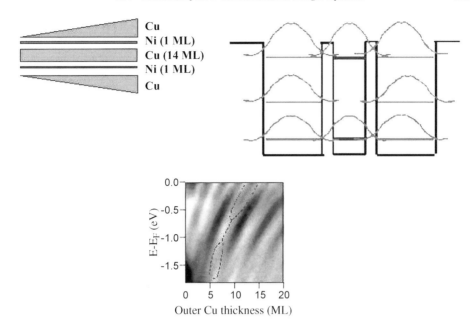

Figure 8.8. Schematic drawings of the wedge sample (left) and QW energy levels in the system Cu/Ni(1 ML)/Cu(14 ML)/Ni(1 ML)/Cu. The wave functions indicate the interaction channel between outer Cu layers through the center Cu(14 ML) film. Also shown is a photoemission image from the Co/Cu multi-layer triple well. Reproduced with kind permission from Y.Z. Wu (private communication). Also published in Ref. [15].

will be observed, compared to what was seen in Figure 8.6. These results explicitly point to the non-free-electron-like behavior of the electronic states that are involved and directional variation of the crystal potential. The non-parabolic nature of the dispersion of E_F vs $\mathbf{k}_\parallel$ has also been demonstrated in Refs. [17] and [18].

8.7. Reduction of the 3-dimensional Schrödinger equation

Although the phase accumulation model, discussed in the previous sections, turned out to be useful in understanding various aspects of QW states, it is far from being rigorous. In various QW systems, it is known to yield somewhat dubious results, especially when dealing with phase changes. There have been numerous attempts to develop more accurate methods, beyond free-electron-like models, and evaluate the effects of the two-dimensional (planar) metallic, periodic structures on the confined states in various devices using different techniques [19]. Some of these have focused on a Fourier space description of the one-particle Schrödinger wave function. Here, we will examine some of the assumptions made in these Fourier as well as real space models about the confined states, the effects of the planar regions, dimensionality on them as well as the importance of spin asymmetry on the spin filtering process [20].

Our model application consists of multilayered slabs of different materials sandwiched together to form a device. For example, the device may contain several layers of Cu sandwiched between two ferromagnetic slabs of Co as in Co/Cu/Co. We will choose the z direction to be perpendicular to these slabs and the (x, y) to be parallel planes consisting of slabs. These can also be labeled longitudinal and transverse directions respectively. There are several simplifications that are usually made in attempts to calculate properties of such structures. When approximations are made to the wave function near an interface, it is a common practice to separate the transverse (x, y) dependence and the longitudinal (z) dependence (see Ref. [19]). Let us carefully examine the conditions on the one-particle potential and the tunneling states that lead to such a description of the problem at hand. An arbitrary eigenstate here can be expressed in the (planar) Bloch form as

$$\Psi_{\mathbf{k}_\parallel}(\mathbf{r}_\parallel, z) = \sum_{\mathbf{G}_\parallel} C_{\mathbf{G}_\parallel}^{\mathbf{k}_\parallel}(z) \exp\{i(\mathbf{k}_\parallel + \mathbf{G}_\parallel) \cdot \mathbf{r}_\parallel\} \tag{8.4}$$

and the one-particle potential $U(x, y, z)$ can be expressed as

$$U(x, y, z) = \sum_{\mathbf{G}_\parallel} V_{\mathbf{G}_\parallel}(z) \exp(i\mathbf{G}_\parallel \cdot \mathbf{r}_\parallel), \tag{8.5}$$

to elucidate the periodic nature of the potential and the Bloch-like form of the wave function in the planar (i.e., parallel) direction. Here $\mathbf{G}_\parallel$ refers to a planar reciprocal lattice vector, while $\mathbf{r}_\parallel = (x, y)$.

It is possible to analyze the effects of the planar states on the perpendicular behavior in several different ways. First, using the Fourier coefficients defined above, the complete one particle solution may be expressed as:

$$-\frac{\hbar^2}{2m}\frac{\partial^2}{\partial z^2} C_{\mathbf{G}_\parallel}^{\mathbf{k}_\parallel}(z) + \sum_{\mathbf{G}_\parallel' \neq \mathbf{G}_\parallel} V_{\mathbf{G}_\parallel - \mathbf{G}_\parallel'}(z) C_{\mathbf{G}_\parallel'}(z)$$

$$= \left\{ E - \frac{\hbar^2}{2m}(\mathbf{k}_\parallel + \mathbf{G}_\parallel)^2 - V_0(z) \right\} C_{\mathbf{G}_\parallel}^{\mathbf{k}_\parallel}(z). \tag{8.6}$$

Note that unless the (parallel and z direction) potential coefficients $V_{\mathbf{G}_\parallel - \mathbf{G}_\parallel'}(z)$ are weak compared to the relevant energy scale of the problem, the above equation couples the Fourier coefficients of the wave function through the potential coefficients and hence does not necessarily yield exponentially decaying solutions in the perpendicular (z) direction for the wave function even if the energy E satisfies the condition

$$E < \frac{\hbar^2}{2m}(\mathbf{k}_\parallel + \mathbf{G}_\parallel)^2 + V_0(z), \tag{8.7}$$

for all values of z.

Consider the wave function as defined in Equation (8.4). An assumption that is usually made [19] when looking for such solutions is the following "separability" condition; i.e.:

$$\Psi = \xi'(x, y)\phi'(z). \tag{8.8}$$

The assumption of the wave function separating into planar (ξ') and perpendicular (ϕ') parts is equivalent to having $C_{\mathbf{G}_\parallel}(z) = \phi'(z)D_{\mathbf{G}_\parallel}$, where $D_{\mathbf{G}_\parallel}$ Fourier coefficient has no z dependence. These ideas can be expressed in terms of the one particle potential

$U(x, y, z)$ and its Fourier expansion given by Equation (8.5). It is clear that if this potential satisfies the (additivity) condition, $U(x, y, z) = U_1(x, y) + U_2(z)$ with no coupling between the planar x, y and perpendicular z dependencies, then the above separation of variables in the wave function can be justified. In such situations, the Hamiltonian H also becomes additive as $H(x, y, z) = H_1(x, y) + H_2(z)$ and a simple quantum well equation,

$$\left[-\frac{\hbar^2}{2m}\frac{\partial^2}{\partial z^2} + U_2(z)\right]\phi'(z) = E_C\phi'(z), \tag{8.9}$$

can be obtained for the $\phi'(z)$.

However, we argue that the above assumptions are too restrictive when one is dealing with states that have p, d or f character in local orbital angular momentum. The coupling of the three directions in the wave functions have to be dealt with more rigor, since such atomic wave functions are unlikely to satisfy the separability condition expressed in Equation (8.8). Significant corrections will be necessary if the separability condition is used as a starting point for better calculations. Hence we have sought a different starting point for carrying out the reduction of the three-dimensional Schrödinger equation. The envelope function approach discussed below provides an ideal and formal platform for handling the situation at hand.

8.8. Envelope functions and the full problem

With the advent of methods related to semiconductor quantum wells, the simplistic theory surrounding them has been so successful that sometimes it is easy to undermine its connections to the well understood regime of weak perturbations in bulk crystals. However, since the potentials employed in quantum wells are strongly perturbed at various boundaries, some formal justification seems necessary in order to use simple quantum well equations for multilayer systems. In fact, some applications of quantum well based techniques to semiconductor heterostructures have been justified using an envelope function approach [21].

In principle, the relevant many particle Hamiltonian in all the different regions of the heterojunction carries all the necessary information and its appropriate eigenstates can be used to describe, for example, quantum tunneling in such a device. However, this problem is highly nontrivial and various approximations are sought in order to simplify it. First principles methods, such as those based on the density functional theory, can be utilized for this purpose, but less complicated approaches that can reduce the computational burden are quite attractive. The envelope function method introduced by Bastard [22] is one such approach. Here the real space equations satisfied by the envelope functions were equivalent to the $\mathbf{k} \cdot \mathbf{p}$ method of Kane [23] with the band edges allowed to be functions of position. For heterojunctions with planar metallic regions, the applications of such ideas can be clarified and presented in a relatively straightforward way, starting from the non-relativistic three-dimensional Schrödinger equation:

$$\left[-\frac{\hbar}{2m}\nabla^2 + U_\sigma(x, y, z)\right]\Psi(x, y, z) = E\Psi(x, y, z). \tag{8.10}$$

As is commonly done for itinerant systems, the spin dependence in the Hamiltonian has been absorbed into a spin dependent potential U_σ, and hence, from now on, we will drop the spin index σ and focus on the reduction of a single, one-particle Schrödinger equation. In the parallel direction, the metallic as well as insulating regions are assumed to have perfect, two-dimensional crystalline order, giving rise to extended electronic states with well defined parallel Bloch momenta $\hbar \mathbf{k}_\parallel$. We also assume perfect (parallel) lattice vector matches at various interfaces.

An important point to note here is that in these problems which involve heterojunctions, there are (at least) two relevant length scales; namely the interatomic length scale and the scale associated with the (confining) structures. The envelope function may vary on the latter length scale or on a combination of the two. Based on this argument, one can expect a nontrivial envelope function, when it exists, to modify the rapidly varying atomic wave functions. Hence, we may express the full problem (ignoring the spin dependence) as

$$
\left(-\frac{\hbar^2}{2m}\frac{\partial^2}{\partial z^2} + \left[-\frac{\hbar^2}{2m}\left(\frac{\partial^2}{\partial x^2} + \frac{\partial^2}{\partial y^2} \right) + U(x, y, z) \right] \right) \xi(x, y, z)\phi(z)
$$
$$
= E\xi(x, y, z)\phi(z). \tag{8.11}
$$

The function $\xi(x, y, z)$ can be thought of as a wave function with rapid variations on the atomic scale that has the two-dimensional Bloch character, while $\phi(z)$ is an envelope function that varies on the length scale of the confining structures or some combination of the two length scales. Note that we do not make an assumption on separability as in Equation (8.8). The existence of a nontrivial envelope, as identified by the above equation, will be used as a prerequisite for the existence of quantum well (or other) confined states. Our search is for envelope functions, as defined above, that are likely to arise due to the confining structures. One can question the validity of such an expression, and similar forms have been suggested [21] such as an expansion using products of Bloch states and envelope functions. Here, we use the above form and associate an eigenvalue $E_\parallel$ with the function $\xi(x, y, z)$ through the following eigenvalue problem.

$$
\left(-\frac{\hbar^2}{2m}\left[\frac{\partial^2}{\partial z^2} + \left(\frac{\partial^2}{\partial x^2} + \frac{\partial^2}{\partial y^2} \right) \right] + U(x, y, z) \right) \xi(x, y, z) = E_\parallel \xi(x, y, z). \tag{8.12}
$$

We note that this is similar to how two-dimensional band structure is calculated for thin films. Although for simplicity, we do not address interface roughness and other similar issues in this paper, the potential $U(x, y, z)$ in Equation (8.11) can be modified to include such effects with appropriate changes in the boundary conditions (and a z dependent potential in Equation (8.14)). With the above definition of $E_\parallel$ (subband energy), the full Schrödinger equation can be used to obtain a differential equation for the envelope function $\phi(z)$ in the following manner.

$$
-\frac{\hbar^2}{2m}\left[\xi(x, y, z)\frac{\partial^2}{\partial z^2}\phi(z) + 2\frac{\partial}{\partial z}\phi(z)\frac{\partial}{\partial z}\xi(x, y, z) \right]
$$
$$
= (E - E_\parallel)\xi(x, y, z)\phi(z). \tag{8.13}
$$

The second term on the left hand side of the above equation contains a product of first order derivatives and describes some coupling of the lateral (i.e., planar) and longitudinal (i.e., perpendicular) coordinates, in addition to any coupling that is already contained in $E_\parallel$. This coupling term can be simplified using averages of $\xi(x, y, z)$ over the planar x, y coordinates. In general, this will result in a z dependent term and a second order differential equation for $\phi(z)$ as

$$-\frac{\hbar^2}{2m}\left[\frac{\partial^2}{\partial z^2}\phi(z) + P_\parallel(z)\frac{\partial}{\partial z}\phi(z)\right] = (E - E_\parallel)\phi(z), \tag{8.14}$$

with

$$P_\parallel(z) = \left(2\iint dx\,dy\,\xi^*\frac{\partial\xi}{\partial z}\right)\bigg/\left(\iint dx\,dy\,\xi^*\xi\right). \tag{8.15}$$

The double integral over the planar coordinates (x, y) has to be carried out over a suitable two-dimensional unit cell. This is a mathematically rigorous result, based solely on the assumptions stated previously, and the theory at this level cannot and should not distinguish between metals, semiconductors or insulators.

Now we can look for possible simplifications to Equation (8.14) by monitoring the properties of $P_\parallel(z)$. Noting that the imaginary part of $P_\parallel(z)$ (Im $P(z)$) is directly related to the (x, y) averaged flux, $J(z)$, along the z direction, we can make use of (steady state) flux conservation which leads to $\partial J(z)/\partial z = 0$; i.e., the conservation of flux implies that Im($P_\parallel$) has to be independent of z. When $P_\parallel(z)$ has a non negligible real part, we cannot make use of the above argument, and the simplifications that follow (i.e., the existence of confined states through Equation (8.17)) will not be necessarily valid. However, note that even for this situation, we have achieved the reduction of the three-dimensional Schrödinger equation to a one-dimensional one. Now setting $P_\parallel(z) = Q_\parallel$ (z-independent), the substitution

$$\phi(z) = \exp(-zQ_\parallel/2)\zeta(z) \tag{8.16}$$

can be used to eliminate the first order derivative leading to a familiar equation, similar to Equation (8.9). The function $\zeta(z)$ appearing here has a different interpretation, as a part of *an envelope function*, and the boundary conditions in the quantum well problem should be applied to the envelope function $\phi(z)$ or $\zeta(z)$. Note that if $\zeta(z)$ and its derivative are continuous across various boundaries along the z direction, then similar properties can be established for $\phi(z)$.

$$-\frac{\hbar^2}{2m}\left[\frac{\partial^2}{\partial z^2}\right]\zeta(z) = \left[E - E_\parallel(\mathbf{k}_\parallel) - \frac{\hbar^2 Q_\parallel^2}{8m}\right]\zeta(z) = E_C\zeta(z). \tag{8.17}$$

The above equation illustrates several important points that are often overlooked or misinterpreted in simplistic quantum well and free electron approaches. As observed in Ref. [19], a possible reduction in tunneling rates can be expected due to the lateral variation of the wave function. The subband structure, i.e., $E_\parallel(\mathbf{k}_\parallel)$, in the multilayers affect the quantum well eigenstates and eigenvalues. The envelope functions are also affected by the averages of the parallel Bloch functions through $Q_\parallel$. For each 'confined energy level' with energy E_C, there exists a continuous subband of states that share the same *perpendicular* wave function $\phi(z)$ but differ in *parallel* Bloch momentum $\mathbf{k}_\parallel$, $E_\parallel$ and $\xi(x, y, z)$. Note that $E_\parallel$ can depend on the thickness of a given multilayer film and carries information not only about the planar structures, but also about the longitudinal coupling, following our definition through Equation (8.12). Finally, the energy E of a given electron in such a quantum mechanical state depends on all of the above.

8.9. Applications – confined states in metallic multilayers

We are now ready to apply our model to test devices. The first device consists of several layers of (nonmagnetic) Cu, sandwiched between two slabs of (ferromagnetic) Co (i.e., Co/Cu/Co) which will be used to discuss spin transmission/rotation effects. For an incident electron of energy E_{total}, we obtain the following electronic perpendicular momenta

$$\hbar k_{\uparrow z} = \sqrt{2m(E_{total} - E_{\parallel}) - \hbar^2 Q_{\parallel}^2/4} \qquad (8.18)$$

and

$$\hbar k_{\downarrow z}^{\Delta} = \sqrt{2m(E_{total} - E_{\parallel} - \Delta) - \hbar^2 Q_{\parallel}^2/4} \qquad (8.19)$$

where $k_{\uparrow}$ ($k_{\downarrow}^{\Delta}$) is the perpendicular wave vector for majority (minority) electrons in the ferromagnetic, metallic regions where the lateral effects have been taken into account using the ideas developed in the previous sections. Here Δ is the spin splitting in the two-dimensional bands assumed to be $\mathbf{k}_{\parallel}$ independent.

In this device, we can rotate the magnetization of the right (R) ferromagnet by an angle θ with respect to the magnetization of the left (L) ferromagnet. The spin rotation is introduced at the boundary $z = 0$ inside the spacer (Cu) layer for simplicity. For this device, solutions to Equation (8.17) in various spatial regions (L and R, subdivided into a, c, d, e) take the form:

$$\begin{bmatrix} \zeta_{a\uparrow}^{L} \\ \zeta_{a\downarrow}^{L} \end{bmatrix} = \begin{bmatrix} Ie^{ik_{\uparrow}^{a}z} + R_{\uparrow\uparrow}e^{-ik_{\uparrow}^{a}z} \\ R_{\uparrow\downarrow}e^{-ik_{\downarrow}^{\Delta a}z} \end{bmatrix}$$

$$\begin{bmatrix} \zeta_{c\uparrow}^{L} \\ \zeta_{c\downarrow}^{L} \end{bmatrix} = \begin{bmatrix} C_{\uparrow+}e^{ik_{\uparrow}^{c}z} + C_{\uparrow-}e^{-ik_{\uparrow}^{c}z} \\ C_{\downarrow+}e^{ik_{\downarrow}^{c}z} + C_{\downarrow-}e^{-ik_{\downarrow}^{c}z} \end{bmatrix}$$

$$\begin{bmatrix} \zeta_{d\uparrow}^{R} \\ \zeta_{d\downarrow}^{R} \end{bmatrix} = \begin{bmatrix} D_{\uparrow+}e^{ik_{\uparrow}^{d}z} + D_{\uparrow-}e^{-ik_{\uparrow}^{d}z} \\ D_{\downarrow+}e^{ik_{\downarrow}^{d}z} + D_{\downarrow-}e^{-ik_{\downarrow}^{d}z} \end{bmatrix}$$

$$\begin{bmatrix} \zeta_{e\uparrow}^{R} \\ \zeta_{e\downarrow}^{R} \end{bmatrix} = \begin{bmatrix} T_{\uparrow\uparrow}e^{ik_{\uparrow}^{e}z} \\ T_{\uparrow\downarrow}e^{ik_{\downarrow}^{\Delta e}z} \end{bmatrix}. \qquad (8.20)$$

Finally, we impose the boundary conditions at $z \pm z_2$ (where $2z_2$ is the thickness of the center slab) and at $z = 0$ for spin rotations.

$$\begin{bmatrix} \zeta_{a\uparrow}^{L}(-z_2) \\ \zeta_{a\downarrow}^{L}(-z_2) \end{bmatrix} = \begin{bmatrix} \zeta_{c\uparrow}^{L}(-z_2) \\ \zeta_{c\downarrow}^{L}(-z_2) \end{bmatrix}$$

$$\begin{bmatrix} \zeta_{d\uparrow}^{R}(0) \\ \zeta_{d\downarrow}^{R}(0) \end{bmatrix} = S(\theta) \begin{bmatrix} \zeta_{c\uparrow}^{L}(0) \\ \zeta_{c\downarrow}^{L}(0) \end{bmatrix}$$

$$\begin{bmatrix} \zeta_{d\uparrow}^{R}(z_2) \\ \zeta_{d\downarrow}^{R}(z_2) \end{bmatrix} = \begin{bmatrix} \zeta_{e\uparrow}^{R}(z_2) \\ \zeta_{e\downarrow}^{R}(z_2) \end{bmatrix} \qquad (8.21)$$

and identically for the first derivative. The matrix

$$S(\theta) = \begin{pmatrix} \cos(\theta/2) & \sin(\theta/2) \\ -\sin(\theta/2) & \cos(\theta/2) \end{pmatrix}$$

is the spinor transformation, where $\theta \neq 0$ is tied to the spin rotation effects as discussed below. We can then fully determine the transmission coefficients utilizing numerical techniques.

8.9.1. *Spin transmission and rotations*

An interesting application of the above ideas has to do with ballistic transport and spin rotation effects. When polarized electrons are transmitted from one region to another with a different polarization axis, they experience a spin-torque and a transfer of angular momentum to the new medium [24] as discussed in Chapter 7. These ideas have now been demonstrated experimentally, for example, through the phenomenon of giant magnetoresistance where large current densities flowing perpendicular to the films have been observed in reversals of magnetization [25], and spin precession [26]. This field is an emerging one related to 'magnetoelectronics' and many new experiments are expected to be conducted on spin transmission/rotation effects in magnetic multilayer systems.

Here we do not wish to focus on the mechanisms of spin transfer torque (discussed in Chapter 7) but simply use the spin rotation angle as an input to our calculations and obtain the corresponding transmission coefficients in a device consisting of two magnetic films separated by a nonmagnetic one (i.e., Co/Cu/Co). Spin rotation effect is introduced, by hand, at the center of the nonmagnetic film (Cu) for simplicity as has been done previously [27]. However, unlike in the previous studies, the underlying band structure and lateral effects have been taken into account by using the theory of confined states developed here. These band dispersions play a crucial role in determining the spin dependent transport properties. For example, when such confined states in a ferromagnetic film are located in a gap of minority spin states, the system can act as an almost perfect spin filter. The $\mathbf{k}$-dependent transmission coefficients, $T_{\uparrow\uparrow}$ and $T_{\uparrow\downarrow}$, that were introduced in the previous section can be used through a Landauer type formula [28],

$$G = \frac{e^2}{h} \sum_{\mathbf{k}_\parallel} T(\mathbf{k}_\parallel), \tag{8.22}$$

to obtain the conductivity in the quantum well problem under discussion. Here we calculate transmission coefficients along $\bar{\Gamma}\bar{X}$ using selected subbands, as defined in Equation (8.23),

$$E_\parallel(\mathbf{k}_\parallel) = E_\parallel^0 + W\{1 - \cos(k_x a)\}, \tag{8.23}$$

in the transmission device (Co/Cu/Co) for the following set of parameters (Figure 8.9) for illustrational purposes: W(Co) = 0.07, W(Cu) = 0.06, $\Delta = 0.19$ (all in Rydbergs), $z_2 = 13.6$, $a = 6.8$ (all in Bohrs).

At a given (total) energy of the incoming electron and a given relative spin orientation of the Co films, transmission coefficients $T_{\uparrow\uparrow}$, $T_{\uparrow\downarrow}$ have been obtained from Equations (8.20) and (8.21) as a function of the parallel Bloch momentum, $\mathbf{k}_\parallel$. From this figure, it is easy to

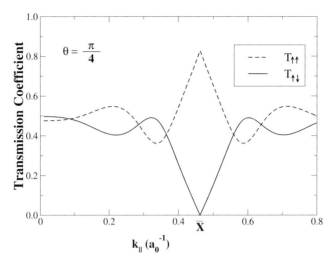

Figure 8.9.   Transmission coefficients $T_{\uparrow\uparrow}$, $T_{\uparrow\downarrow}$ as a function of $\mathbf{k}_\parallel$ along $\bar{\Gamma}\bar{X}$ in the device (Co/Cu/Co) mentioned in the text for a relative spin orientation $\theta = \frac{\pi}{4}$. At the zone boundary $\bar{X}$, the device filters out the 'minority' spins. Away from the zone boundary both 'majority' and 'minority' spins are transmitted.

see the effects of different types of Bloch states on the tunneling. Near the zone boundary $\bar{X}$, for the given relative spin orientation θ, we see the device filtering out "minority" spins, while at other $\mathbf{k}$-points along $\bar{\Gamma}\bar{X}$, a mixture of both "majority" and "minority" spins are transmitted. This is a direct result of the upward dispersion of the selected subbands along $\bar{\Gamma}\bar{X}$ and the spin splitting in Co, pushing the minority band closer to the given (total) energy of the electron.

8.9.2.   Angle resolved photoemission and inverse photoemission

Angle Resolved Photoemission (ARP) techniques have become a useful tool for studying electronic states and their (band) dispersions in structures with two-dimensional (planar) Bloch symmetry. As discussed in Chapter 5, in ARP experiments photons are used to eject electrons from occupied states while in inverse ARP, photons are ejected, when the above ARP process is reversed. Photoemission is a many-electron phenomenon and an evaluation of the spectral function and appropriate self-energies yield the full photoemission spectrum and peak widths. However, here we follow a simpler approach and focus on one particle energy states and the changes introduced by the multilayering. In this approach, the energy E that is measured for the photoemitted electron can be associated with the full Schrödinger equation (8.10). Existence of an envelope function as defined in Equation (8.11) and satisfying the imposed boundary conditions are necessary for confined states. The energy $E_\parallel$ as defined through Equation (8.12) has to do with the ordinary subband states due to the periodic potential in the planar directions. If the multilayering does not play any role for a given energy E at a given $\mathbf{k}_\parallel$, then $E_\parallel$ and E have to be identical when these solutions exist in Equations (8.10) and (8.12). In such cases, $\phi(z)$ turns out to be a simple multiplicative constant and Equation (8.13) is consistent with this scenario. However, when confined states exist, they can, in general, alter the usual dispersions $E_\parallel$

that are observed in their absence. According to the theory developed above, confined states can be identified as states for which solutions to the differential equation (8.17) exist for a given energy E of the electron. It is also important to realize that, for a given energy E, these states may not exist for all values of $\mathbf{k}_\parallel$ along a given direction of the 2 dimensional Brillouin zone. We can also derive an expression for the dispersions of the energy bands that can be associated with such states.

In general, the dispersion of the total energy of the electron E, i.e., the energy of the bound electron as a function of $\mathbf{k}_\parallel$, is found whenever confined states exist with appropriate $\zeta(z)$ as solutions to Equation (8.17). In the present context, with boundary conditions appropriate to stationary states (determined by the film geometry), we obtain a set of discrete energies (E_n) and states for the one-dimensional quantum well or barrier problem where

$$E = E_n + E_\parallel(\mathbf{k}_\parallel) + \frac{\hbar^2 Q_\parallel^2}{8m}. \tag{8.24}$$

Hence, the observed energies E in an ARP experiment will depend on the existence of these discrete E_ns and the corresponding ζ functions (as defined in Equation (8.17)) satisfying the relevant boundary conditions. As in simple quantum well problems, solutions of odd and even parity may sometimes be associated with E_n and these will depend on the film geometry, interfaces, and growth conditions.

8.10. Summary

Confined geometries give rise to confined states in materials systems. Metallic multilayers provide a well studied example where such confinement and the existence of QW states have been clearly established. Experimentally, these confined QW states and their properties have been analyzed for numerous multilayer systems. Photoemission experiments that have been quite useful in this regard were discussed in some detail. Theoretically, a simple model (PAM), that takes into account phase shifts, has provided numerous insights into the existence of QW states. A more detailed envelope function approach, designed to reduce the three-dimensional Schrödinger equation to a one-dimensional one and analyze the lateral effects of metallic multilayer films through a hypothetical subband structure that varies with the two-dimensional Bloch vector, was also discussed. Ballistic transmission in a magnetoelectronic device consisting of Co/Cu/Co has been examined and shows a clear dependence on the Bloch vector $\mathbf{k}_\parallel$. The nonmagnetic spacer plays a key role in determining the periods associated with various resonances crossing the Fermi level. Also, the method of reduction of the 3-dimensional Schrödinger equation with appropriate modifications may be used in other confined geometries such as nanoparticles and wires.

Acknowledgements

The author is grateful to Drs. Z.Q. Qiu and Y.Z. Wu for granting permission to use material from their work, in addition to helpful discussions.

References

 [1] J.E. Ortega, F.J. Himpsel, Phys. Rev. Lett. **69**, 844 (1992).
 [2] J.E. Ortega, F.J. Himpsel, G.E. Mankey, R.F. Willis, Phys. Rev. B **47**, 1540 (1993).
 [3] Physics Today **48**, 24–63 (1995).
 [4] P. Segovia, E.G. Michel, J.E. Ortega, Phys. Rev. Lett. **77**, 3455 (1996).
 [5] F.G. Curti, A. Danese, R.A. Bartynski, Phys. Rev. Lett. **80**, 2213 (1998).
 [6] R.K. Kawakami, et al., Phys. Rev. Lett. **80**, 2213 (1998).
 [7] R.K. Kawakami, et al., Nature **398**, 132 (1999).
 [8] A. Danese, D.A. Arena, R.A. Bartynski, Prog. Surf. Sci. **67**, 249 (2001).
 [9] Z.Q. Qiu, N.V. Smith, J. Phys.: Condens. Matter **14**, R169–R193 (2002).
[10] N.V. Smith, Phys. Rev. B **32**, 3549 (1985).
[11] N.V. Smith, N.B. Brookes, Y. Chang, P.D. Johnson, Phys. Rev. B **49**, 332 (1994).
[12] A. Danese, R.A. Bartynsky, Phys. Rev. B **65**, 174419 (2002).
[13] K.G. Garrison, Y. Chang, P.D. Johnson, Phys. Rev. Lett. **71**, 2701 (1993).
[14] C. Carbone, E. Vescovo, O. Rader, W. Gudat, W. Eberhardt, Phys. Rev. Lett. **71**, 2805 (1993);
     S. Yang, et al., Phys. Rev. Lett. **70**, 849 (1993).
[15] Y.Z. Wu, C. Won, E. Rotenberg, H.W. Zhao, Qi-Kun Xue, W. Kim, T.L. Owens, N.V. Smith, Z.Q. Qiu, Phys. Rev. B **73**, 125333 (2006).
[16] E. Rotenberg, Y.Z. Wu, J.M. An, M.A. Van Hove, A. Canning, L.W. Wang, Z.Q. Qiu, Phys. Rev. B **73**, 075426 (2006).
[17] Y.Z. Wu, C.Y. Won, E. Rotenberg, H.W. Zhao, F. Toyoma, N.V. Smith, Z.Q. Qiu, Phys. Rev. B **66**, 245418 (2002).
[18] Y.Z. Wu, Private communication.
[19] W.H. Butler, X.-G. Zhang, T.C. Shulthess, J.M. MacLaren, Phys. Rev. B **63**, 092402 (2001).
[20] C. Abbott, G.W. Fernando, M. Rasamny, Phys. Rev. B **69**, 205412 (2004).
[21] M.G. Burt, J. Phys. Condens. Matt. Phys. R **53**, (1999);
     M.G. Burt, J. Phys. Condens. Matt. Phys. R **4**, 6651 (1991).
[22] G. Bastard, Phys. Rev. B **24**, 5693 (1981);
     G. Bastard, J.A. Brum, R. Ferreira, in: Solid State Physics, vol. 44, Academic Press, New York, p. 229 (1991).
[23] E.O. Kane, in: Semiconductors and Semimetals, vol. 1, Academic Press, New York, p. 75 (1975).
[24] J.C. Slonczewski, J. Magn. Magn. Mater. **159**, L1 (1996);
     J.C. Slonczewski, J. Magn. Magn. Mater. **195**, L261 (1999);
     L. Berger, Phys. Rev. B **54**, 9353 (1996).
[25] E.B. Myers, D.C. Ralph, J.A. Katine, R.N. Louie, R.A. Buhrman, Science **285**, 867 (1999).
[26] W. Weber, S. Rieseen, H.C. Siegmann, Science **291**, 1015 (2001).
[27] J.C. Slonczewski, Phys. Rev. B **39**, 6995 (1989).
[28] R. Landauer, Philos. Mag. **21**, 863 (1970).

Chapter 9

Half-Metallic Systems: Complete Asymmetry in Spin Transport

9.1. Introduction

The major technological breakthroughs we have encountered in this volume can be directly tied to creative ways of making new materials such as, controlling the growth and coherently forming layered systems which are magnetic and nonmagnetic. Without such synthesis, fabrication and analysis, none of the technological advances discussed here would have been possible. When one looks at the elemental or other basic components of some of these technical devices, they are all more or less "well-known" materials. The important critical advances occurred only when it was recognized how to put things together with a firm hold on the underlying theoretical concepts. Even if the initial attempts were Edisonian trial and error type, the rapid developments we have seen over the past few decades provide ample testimony to numerous branches of materials science coalescing and blossoming into one fast growing unit. There are some special classes of the "well-known" materials that are likely to play a crucial, resurgent role in these endeavors. Half-metallic alloys, to be discussed in this chapter, constitute one such class of materials.

When discussing metallic transport, one usually encounters systems with metallic spin-up and spin-down electrons, i.e., both spins contribute to the Fermi level density of states (and transport). In the search for highly spin polarized systems, one could envision having a situation where one spin channel shows metallicity while the other spins are completely insulating. Such materials, where tunneling currents show 100% spin polarization when used in tunneling devices, have been found and are called half-metals. Half-metals have unusual electronic properties. For one type of spin, the half-metal looks metallic with a well defined Fermi surface. However, for the opposite spin, there is a gap in the state density (see Figure 9.1). This is different from normal, magnetic metals, with an almost filled majority spin d states such as Ni, which may have mostly minority spin states at the Fermi level. However, in such metals s and p states give rise to some none-zero density of states, consisting of both up and down spins, at the Fermi level. In half-metals, there is a clear gap in the density of states corresponding to one type of spin. These half-metals do conduct electricity but do not show a high field magnetic susceptibility. All half-metals contain more than one element with most of them being oxides or Heusler type alloys (see Tables 9.1 and 9.2).

Although the exploration of real half-metallic compounds as possible components in spintronic devices is relatively new, De Groot and Buschow [1] introduced the concept of half-metallicity in the 1980s to explain the semi-insulating behavior of compounds such as

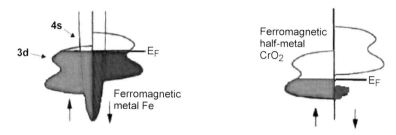

Figure 9.1.   Schematic comparison of the density of states of a ferromagnetic half-metal with that of ferromagnetic Fe. Note that in the half-metal shown here the density of minority spin states at the Fermi level is zero, while for the ferromagnetic metal, there are both majority and minority states at the Fermi level.

Table 9.1   Typical examples of half-metals

Crystal structures	Examples	References
Perovskite and double perovskite	$La_{0.7}Sr_{0.3}MnO_3$, Sr_2FeMoO_6	[7,6]
Rutile	CrO_2	[5]
Spinel	Fe_3O_4	[8]
Half Heusler $C1_b$	NiMnSb	[1]
Full Heusler $L2_1$	$Co_2Cr_{1-x}Fe_xAl$	[15]

Table 9.2   Comparison of selected half-metals

Density of States	Conductivity	$\uparrow$ DOS at E_F	$\downarrow$ DOS at E_F	Examples
half-metal	metallic	itinerant	none	CrO_2, NiMnSb
half-metal	metallic	none	itinerant	Sr_2FeMoO_6
half-metal	nonmetallic	none	localized	$LaMnO_3$

Characterization of DOS for selected half-metals (from Ref. [10]).

NiMnSb and PtMnSb. These are categorized as semi-Heusler alloys which are different from the parent Heusler alloy due to a vacant site in the unit cell as discussed below. Heusler alloys, which are magnetic, also show similar behavior, as concluded by Kubler *et al.* in Ref. [2] where it was seen that the localized character of the magnetism was due to the exclusion of the minority *d* spins from Mn. This property imposes restrictions on the possible low energy spin excitations.

Heusler alloys are promising candidates for spintronic applications (for example in TMR devices) discussed in this volume. They probably have the greatest potential for such applications at room temperature due to high Curie temperatures and close lattice matchings with III–V semiconductors. A number of these Heusler alloys show half-metallic behavior [3]. There are two distinct groups of Heusler alloys related in the zincblende structure, one of which has the form XYZ and crystallizes in the $C1_b$ structure with a fcc sublattice occupied by the three atoms X, Y, and Z and a vacant site (see Figures 9.2 and 9.3). These are sometimes referred to as half Heusler alloys. The full Heusler alloys

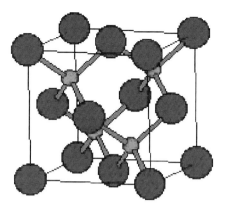

Figure 9.2. Zincblende structure.

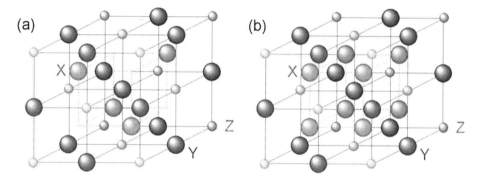

Figure 9.3. Half-Huesler (XYZ) and full Heusler (X$_2$YZ) alloy structures mentioned in the text.

have the form X$_2$YZ with the vacant site in XYZ occupied by X in the L2$_1$ structure. The half Heusler alloys were identified early on for their half-metallicity; at that time, it was not experimentally known that the full Heusler ones have this property although there were first principles predictions to that effect.

In real half-metals, conduction is carried out almost entirely by one spin channel. Unlike in superconductors, metals or insulators, there is no clear experimental signal to identify half-metals. If the minority (or majority) gap is a true gap (with zero total density of states) then it is possible to conclude that the majority and minority counts of spin, $n^{\uparrow}$ and $n^{\downarrow}$, have to be integers per unit cell. This is because that the sum $n^{\uparrow} + n^{\downarrow}$ is an integer per unit cell. (Alternatively, a given band carries an integral number of electrons of one spin type per unit cell and hence a gap in one spin channel requires an integral count of the corresponding spin per unit cell.) Due to the existence of a gap in one spin channel, the corresponding spin count must also be an integer which implies that the other spin count is also an integer.

Electronic structure calculations have been valuable in the search for such half-metals since experimental searches can be tedious and identification of half-metals is somewhat complicated. Since the band gap for one spin type is the defining feature of these materials, it is important to examine the nature and origin of these gaps. There are different ways to categorize half-metals and here we list three different categories of gaps that can be identified following Ref. [4]. These are: (a) covalent band gaps, (b) charge-transfer gaps, and (c) $d - d$ band gaps. Half-metals in different categories listed above respond differently to external parameters such as pressure. This is due to different electronic structures that give rise to half-metallicity in these 3 categories [4].

9.1.1.   Covalent gaps

Familiar semiconductors of group III–V, such as GaAs, are strongly related to these half-metals. A well-known example is NiMnSb which has the crystal structure C1$_b$, closely related to zincblende structure [1]. (See Figures 9.2 and 9.3.) The crystal structure and site coordination are quite important for half-metallicity. The band structure and the nature of bonding for the semiconducting spin channel are quite similar to the semiconductors of group III–V. For the metallic spin channel, wide bands are found with effective mass being essentially that of a free electron. Tetrahedral coordination is found for Mn and Sb in NiMnSb mentioned above. These materials are soft magnets; i.e., their magnetic behavior is similar to iron with the presence of minority states being essential.

9.1.2.   Charge-transfer gaps

Charge-transfer gaps are found in strongly magnetic half-metallic compounds. Here the d band of the transition metal is empty for the minority spins while the $s - p$ bands are localized at the anions. The crystal structure does not play a strong role in the formation of the minority spin gap. These behave more or less as hard magnets, i.e., a hypothetical increase in the exchange-splitting does not lead to increases in the magnetic moment; this could be due to majority d bands being completely full or minority ones that are empty. Examples of such half-metals are CrO$_2$ [5], colossal magnetoresistance (CMR) materials including the double-perovskites [6].

9.1.3.   d-d band gap materials

These materials consist of narrow d bands which give rise to gaps due to crystal-field splitting. Fermi level could fall into a gap of one type of spin due to exchange splitting. Some examples are Fe$_3$O$_4$ [8] and Mn$_2$VAl [9]. The magnetism in these compounds is not that strong (i.e., these are soft magnets).

9.2.   Half Heusler alloys: NiMnSb and PtMnSb

As stated previously, NiMnSb exhibits half-metallic behavior in addition to some novel properties. PtMnSb also shares the half-metallicity. In these compounds the Fermi level falls in a gap of the minority spin states and decouples the two spins in the ordered state. The metallicity is entirely due to the majority spins at the Fermi level. This property leads to a huge Kerr effect in PtMnSb, which at the time of discovery was the largest among metallic systems [17]. Magnetization and resistivity of single crystals of this half-metal seems to show a transition at low temperature, from a Heisenberg type magnet to an itinerant one [11]. A Stoner type mechanism is said to take over at higher temperatures. These conclusions are based on the following observations in Ref. [11]: (a) the existence of a small magnetocrystalline anisotropy, (b) the presence of a transition in magnetization and resistivity around 70 K. Below this temperature, magnetization varies as $T^{3/2}$ satisfying a classical spin wave law for Heisenberg magnets. The resistivity shows a T^2 behavior below 50 K as in Heisenberg magnets and strongly interacting Fermi liquids but increases rapidly around 70 K and flattens above this temperature, indicating some change or transition in the electronic properties.

9.3.   Full Heusler alloys: Co_2MnSi, Co_2MnGe, $Co_2Cr_{1-x}Fe_xAl$

There have been extensive efforts in searching for half-metallicity in full Heusler alloys (see, for example, Refs. [12–14]). Recently, Inotoma *et al.* were able to demonstrate half-metallicity in several Heusler alloys [15,16]. A noteworthy feature of the Heusler alloys, such as Co_2MnSi and Co_2MnGe, is the high degree of spin polarization at room temperature. This is obviously important when fabricating spintronic devices that operate at room temperature. Figure 9.4 also shows magnetization curves for Co_2MnSi and Co_2MnGe films illustrating the room (and higher) temperature magnetic hysteresis. In addition, for thin films of these two Heusler compounds studied in Ref. [15], Co_2MnSi and Co_2MnGe, several differences in physical properties have been observed when compared with their bulk counterparts.

Although the bulk crystal structure for Heusler alloys is L2₁, thin films of these alloys may crystallize in a different structure. For example, $Co_2Cr_{0.6}Fe_{0.4}Al$ films grown on thermally oxidized Si substrates at room temperature do not show the L2₁ structure but crystallizes in the B2 structure. In fact, depending on the Fe content in $Co_2Cr_{1-x}Fe_xAl$ (CCFA), its thin films can form in either B2 or A2 structures [16]. As before, note the high degree of spin polarization at room temperature as demonstrated in Figure 9.5. In addition, these researchers were able to fabricate magnetic tunnel junctions (MTJs) consisting

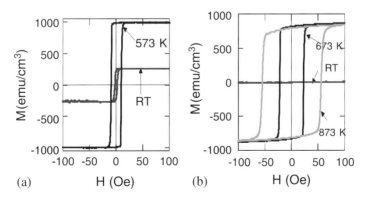

Figure 9.4.   Magnetic hysteresis curves for Co_2MnSi and Co_2MnGe films [16]. Reproduced with kind permission from Elsevier.

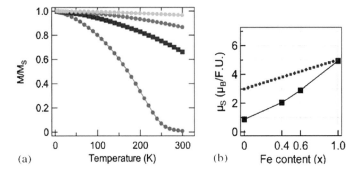

(a)   (b)

Figure 9.5.   (a) Temperature dependence of the magnetization of $Co_2Cr_{1-x}Fe_xAl$ films for $x = 1.0, 0.6, 0.4$ and 0.0 from the top respectively, (b) Fe concentration dependence of the magnetic moment per formula unit at 5 K. The dotted line is for theoretical values [16]. Reproduced with kind permission from Elsevier.

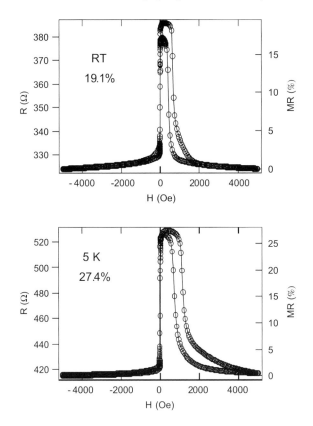

Figure 9.6.   Tunneling magnetoresistance of ferromagnetic half-metal CCFA. Reproduced with kind permission from Elsevier.

of CCFA, an aluminum oxide, and other metallic alloys for various Fe concentrations in CCFA. The maximum tunneling magnetoresistance (TMR) at 5 K was observed at $x = 0.4$ to be about 27% and about 19% at room temperature (see Figure 9.6).

9.4. Chromium dioxide

Half-metallicity in CrO_2 [5] at low temperature can be understood in a rather straight-forward way. It is thought to be arising from the fact that its exchange splitting, which is greater than the valence bandwidth of the majority electrons, is pushing the minority electrons states above the Fermi level (see example in Figure 9.1). Chromium dioxide is unique in the sense that it is the only stoichiometric binary oxide that is a ferromagnetic metal. It is also one of well studied, simple compounds that is a half-metal. Although metastable under ambient conditions, it has a narrow range of stability near 300 °C including a high oxygen pressure region. The crystal structure of CrO_2 is Rutile with each oxygen having three chromium neighbors; each chromium is octahedrally coordinated by oxygen with two short and four long equatorial bonds. The Cr d orbitals are split by the crystal field into three t_{2g} and two e_g orbitals. The t_{2g} orbitals are further split in a nonbonding d_{xy} which lies in the equatorial plane of the octahedron and a d_{xz}, d_{yz} doublet. Electronic structure calculations show that the Fermi level lies in the half-full majority d_{xz}, d_{yz} band (i.e., clearly metallic) while there is a gap of more that 1 eV for the minority spin states at the Fermi level.

9.5. Perovskites and double-perovskites

There are doped colossal magnetoresistive (CMR) perovskites that show half-metallicity. CMR materials are mostly insulators that exhibit large changes in magnetoresistance in response to a magnetic field. They are characteristic of a class of Mn perovskites with a high degree of spin polarization and hence possible candidates for spintronics appli-

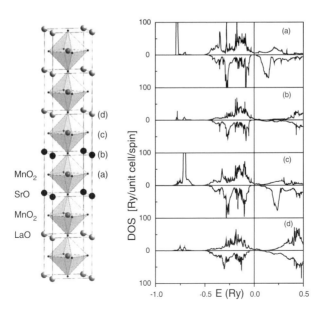

Figure 9.7. Crystal structure and calculated DOS of LSMO (from Ref. [18]).

cations. In spin-resolved photoemission experiments, carried out for the ferromagnetic perovskite $La_{0.7}Sr_{0.3}MnO_3$ (LSMO – see Figure 9.7), half-metallic behavior has been clearly observed well below its Curie temperature [7]. For the majority spin, the photoemission spectrum showed a metallic Fermi level cut-off while for the minority spin, an insulating gap was observed.

In the double-perovskite Sr_2FeMoO_6, first principles calculations show the existence of a gap in the majority states while the minority bands are said to be responsible for the metallic behavior [6]. The significance of the above double-perovskite is its high Curie temperature (around 415 K) which makes it possible to use in tunneling magnetoresistive devices at room temperature.

9.6.   Multilayers of zincblende half-metals with semiconductors

Mavropoulos *et al.* [19] have carried out an interesting set of first principles calculations examining multilayers of zincblende half-metals (namely CrAs and CrSb) with III–V and II–VI semiconductors. Their focus was on the question whether half-metallicity would be conserved at the interface. They found that this can be achieved with not too restrictive conditions. One of these conditions is that the coordination of the transition metal does not change at the interface; i.e., it should have four sp neighbors of electronegative character as in the bulk. As a result, the bonding-antibonding separation (splitting) of the $p - d$ hybrids is retained and half-metallicity is conserved at the interface. Another finding of this study is that all four neighbors need not be of the same type. An examination of the magnetic moment at the interface yielded a non-integer value for the semiconductor/half-metal interfaces studied and a simple sum rule for the moment at the interface was derived. They also concluded that zincblende half-metallic compounds have various desirable qualities for applications in spintronics, such as half-metallic ferromagnetism, high Curie temperatures, coherent growth and half-metallic interfaces with semiconductors.

For more information on half-metallic alloys, the reader is referred to Ref. [20].

References

[1] R.A. de Groot, F.M. Mueller, P.G. van Engen, K.H.J. Buschow, Phys. Rev. Lett. **50**, 2024 (1983).
[2] J. Kubler, A.R. Williams, C.B. Sommers, Phys. Rev. B **28**, 1745 (1983).
[3] W.E. Pickett, J.S. Moodera, Physics Today **39**, 54 (2001);
    I. Galanakis, et al., Phys. Rev. B **66**, 174429 (2002).
[4] C.M. Fang, G.A. de Wijs, R.A. de Groot, J. Appl. Phys. **91**, 8340 (2002).
[5] K. Schwarz, J. Phys. F: Met. Phys. **16**, L211 (1986).
[6] K.L. Kobayashi, et al., Nature **395**, 677 (1998).
[7] J.H. Park, et al., Nature **392**, 794 (1998).
[8] A. Yanase, H. Sitarori, J. Phys. Soc. Jpn. **53**, 312 (1984).
[9] R. Weht, W.E. Pickett, Phys. Rev. B **60**, 13006 (1999).
[10] J.M.D. Coey, M. Venkatesan, J. Appl. Phys. **91**, 8345 (2002).
[11] C. Hordequin, J. Pierre, R. Currat, J. Magn. Magn. Mater. **162**, 75 (1996).
[12] T. Ambrose, J.J. Krebs, G.A. Prinz, Appl. Phys. Lett. **76**, 3280 (2000).

[13] U. Geierbach, A. Bergmann, K. Westerhold, J. Magn. Magn. Mater. **240**, 545 (2002).

[14] M.P. Raphael, et al., Phys. Rev. B **66**, 104429 (2002).

[15] K. Inomata, S. Okumara, R. Goto, N. Tezuka, Jpn. J. Appl. Phys. **42**, L419 (2002).

[16] K. Inomata, S. Okamura, N. Tezuka, J. Magn. Magn. Mater. **282**, 269 (2004).

[17] R.A. de Groot, K.H.J. Buschow, J. Magn. Magn. Mater. **54–57**, 1377 (1986).

[18] C. Banach, R. Tyer, W.M. Temmerman, J. Magn. Magn. Mater. **272–276**, 1963 (2004).

[19] Ph. Mavropoulos, I. Galanakis, P.H. Dederichs, J. Phys.: Condens. Matter **16**, 4261 (2004).

[20] Galanakis, P.H. Dederichs (Eds.), Half-Metallic Alloys, Springer, Berlin (2005).

Chapter 10

Exact Theoretical Studies of Small Hubbard Clusters

10.1. Introduction

During the past few decades, remarkable progress has been made in the "controlled growth" of artificial materials at the nanoscale. Nanoscience has become one of the main frontiers of modern physics and the materials frontier is proving to be the biggest single means to advance our knowledge of nanoscience. Nanomaterials, such as quantum dots, rods, grains, have already found their way into numerous applications such as miniature sensors and are becoming extremely useful in medical and engineering applications. These developments have clearly generated a tremendous interest among both the scientific and nonscientific communities.

The ability to synthesize small clusters containing a few atoms provides a unique opportunity to examine and tune fundamental physical and chemical properties. Hence, it is important to conduct theoretical studies of such clusters. Analytical results pertaining to such systems are understandably rare, but numerical methods can provide significant insight provided sufficiently accurate methods are developed. This chapter describes a number of surprising and novel results obtained using a combination of exact diagonalization of a many-body Hamiltonian and statistical mechanics [1–3].

Understanding the effects of electron correlations and pseudogap phenomena in the (layered) cuprate superconductors comprising of many different phases is regarded as one of the most challenging problems in condensed matter. Although the experimental determination of various inhomogeneous phases in cuprates is still somewhat controversial [4], the underdoped high T_c superconductors (HTSCs) are often characterized by crossover temperatures below which excitation pseudogaps in the normal-state are seen to develop [5]. In the optimally doped cuprates the correlation length of dynamical spin fluctuations is very small [6]. Therefore, microscopic studies, which account accurately for short-range dynamical correlations in finite clusters, are relevant and useful with regard to understanding the physics of the HTSCs. Such studies are also directly relevant to nanomagnetism as well as spintronics applications.

There are several advantages to developing such a theory of materials from the bottom-up, using nanoclusters. Such cluster studies can be used as a guide to constructing the next generation of miniaturized electronics and photonic devices with greater flexibility. These proposed studies can open up new avenues towards potential electronic applications and synthesis of high-quality predictable nanoclusters and self assembled materials than with the current synthetic routes. The exact quantum statistical calculations of electron correlations and thermal properties represent, by definition, the most accurate theory for studies of nanocluster thermodynamics at nonzero temperatures. The geometry of an elementary building block (cluster), its orientation, shape and position in two and three dimensions can be well controlled in a multidimensional parameter space of interaction strength, magnetic field, electron concentration and temperature.

In a recent review article on manganites (Ref. [7]) it is concluded that inhomogeneous 'clustered' states should be considered as a *new paradigm* in condensed matter physics. Cluster studies are becoming more and more relevant due to nanoscale inhomogeneities observed in several important classes of materials. These clusters can also be viewed as prototypical models for monitoring the evolution of bulk electronic structure of materials and transition-metal compounds, such as manganites or HTSC cuprates, starting from the nanometer scale. In addition, independent clusters of various topologies embedded in materials can alter magnetic and other intrinsic properties of materials. The method developed here can be used to describe magnetic instabilities in the spin-tetrahedral systems in terms of weakly coupled tetrahedrons with a singlet-triplet gap and low-lying singlets (to be published). Interestingly, the degenerate electrons in frustrated clusters with a MH-like insulating state exhibit a strong tendency toward superparamagnetism.

Our (thermodynamical) theory provides a mechanism of phase separation into carrier-rich and carrier-poor regions observed in the HTSC cuprates; it has been able to predict some of the recent observations related to local electron pairing on the atomic scale (Ref. [8]), inhomogeneous phase separation (PS) under doping or a magnetic field in HTSCs [30]. These ideas may also be linked to experiments in Carbon Nanotubes (CNs), where a purely electronic mechanism may be responsible for superconductivity. Besides the HTSCs, the PS inhomogeneities may be relevant to doped manganites, fullerenes and CNs with (possible) purely electronic mechanisms of superconductivity (SC). Under appropriate electron doping, we have seen how the homogeneous state becomes unstable towards the formation of small ferromagnetic droplets inside an antiferromagnetic insulating matrix.

The following questions are central to the present study: (i) When treated exactly, what essential features can the simple Hubbard clusters capture, that are in common with the HTSCs as well as other materials where local correlations are strong? (ii) Using simple cluster studies, is it possible to obtain a mesoscopic understanding of electron–hole/electron–electron pairing and identify various possible phases and crossover temperatures? (iii) Do these *nanoclusters* retain important features that are known for large thermodynamic systems?

The significance of the present study is two-fold: First, although numerous properties, including eigenvalues and susceptibilities, of Hubbard clusters have been calculated (for example, see Refs. [9–18]) previously, many open questions remain with regard to microscopic origins of charge-spin separation, pseudogap behavior, and various scenarios of possible pairings at low temperature. Apparently, the above studies did not search for transitions, as we have done here, over an extended parameter space that includes variations in chemical potential, magnetic field, Coulomb repulsion and temperature. Novel insight into the properties of Hubbard clusters is gained by monitoring weak singularities in susceptibilities over this extended parameter space [19]. It is our hope that this work would motivate further studies of electronically driven mechanisms for pairing etc. in correlated nanosystems.

Second, these results may be relevant to doped fullerenes [20], superconducting Carbon nanotubes (CNs) [21], organic superconductors [22] as well as high temperature superconductors (HTSCs), when searching for a purely electronic mechanism for superconductivity. It is imperative that such similarities be carefully examined and understood, given the low dimensionality of CNs and the inhomogeneities observed in the HTSCs [23,24]. In addition, while one-dimensional theorems preclude long-range order, it is still an open question whether finite clusters can retain ground state, short-range correlations at finite temperature [25,3].

Since the discovery of the high temperature superconductors (HTSCs), there has been an intense debate about a possible electron (or hole) pairing mechanism. Early on, Anderson [26] suggested that the large positive onsite Coulomb interaction in the Hubbard model should contain the key to some of the perplexing physics observed in the HTSCs. The ground state of the model, in different geometries, has also been suggested as providing a mechanism for electron pairing in fullerenes and certain mesoscopic structures due to intramolecular correlation effects at moderate U [23]. Our recent work [2] indicates that an ensemble of Hubbard clusters, when connected to a particle reservoir and a thermal bath, possess a vivid variety of interesting thermal and ground state properties. These inferences were drawn by carrying out exact diagonalizations of the many-body Hamiltonian, and using these eigenvalues in a statistical ensemble to study ground state transitions and thermal crossovers, by monitoring susceptibilities, i.e., *fluctuations*. Such ensemble averages have been demonstrated to be relevant for nanoclusters [27]. Although there have been numerous studies of phase separation and similar phenomena in Hubbard systems [28,23], in our opinion, thermal properties and phase instabilities of small Hubbard clusters *at an arbitrary filling* have not been fully explored.

10.2. Methodology and key results

The single orbital Hubbard Hamiltonian,

$$H = -t \sum_{\langle ij,\sigma \rangle} c_{i\sigma}^{+} c_{j\sigma} + \sum_{i} U n_{i\uparrow} n_{i\downarrow}, \tag{10.1}$$

with hopping t and on-site interaction U, has been used in this work with periodic boundary conditions. In order to study thermal properties, the many-body eigenvalues of the Hubbard clusters are combined with the grand canonical potential Ω_U for interacting electrons with

$$\Omega_U = -T \ln \sum_{n \leqslant N_H} e^{-\frac{E_n - \mu N_n - h s_n^z}{T}}, \tag{10.2}$$

where N_n and s_n^z are the number of particles and the projection of spin in the nth state respectively. The grand partition function Z (where the number of electrons N and the projection of spin s^z can fluctuate) and its derivatives are calculated exactly without taking the thermodynamic limit. The response functions related to electron or hole doping (i.e., chemical potential μ) or magnetic field h demonstrate clearly observable, prominent peaks paving the way for strict definitions of Mott–Hubbard (MH), antiferromagnetic (AF), spin pseudogaps and related crossover temperatures [2,3]. Unless otherwise stated, all the energies reported here are measured in units of t (i.e., t has been set to 1 in most of the work that follows). For a complete description of the method, see Refs. [1,2].

It has been shown that in such finite systems, one can define the gap parameters for various transitions and identify corresponding phase boundaries by monitoring maxima and minima in charge and spin susceptibilities [2]. As synthesis techniques improve at a rapid rate, it has become possible to synthesize isolated clusters and hence it is clear that we need not always look at the thermodynamic limit. Finite, mesoscopic structures in suitable topological forms will be realistic enough to synthesize (such as the CNs [21]) and extract fascinating physics. Also, since the HTSCs are known to consist of (stripes and possibly other) inhomogeneities [24], it is possible that these cluster studies may be able to capture some of the essential physics of the HTSCs. The following is a list of key results from these exact (4-site and 8-site) Hubbard cluster studies:

1. Phase diagrams in a temperature-chemical potential (doping) plane and the presence of a multitude of fascinating phases, including Mott–Hubbard like paramagnetic and antiferromagnetic phases [2].
2. Vanishing of a charge gap at a critical set of parameters and thereby providing an effective attraction leading to onset of electron pairing at a critical temperature T_c^P.
3. Spin pairing at a lower temperature (T_s^P) and hence the formation of rigidly bound spin pairs in a narrow, critical region of doping.
4. Low temperature specific heat peak, reminiscent of the experimental, low temperature specific heat behavior in the HTSCs [4].
5. Temperature vs. U phase diagram, indicating the pressure effect on the superconducting transition temperature similar to reported results of recent experiments [20,22,29].
6. The presence of a dormant magnetic state, lurking in the above narrow, critical region of doping, that could be stabilized by either applying a magnetic field, going above the spin pairing temperature, or changing the chemical potential, similar to what is observed in a recent, notable experiment [30].
7. The opening of a pseudogap above the pairing temperature, similar to what is seen in NMR experiments, in both hole and electron doped cuprates [31].
8. Larger clusters with different topologies and higher dimensionality illustrating how the above properties get scaled with size.

10.3.   Charge and spin pairings

These exact studies of 4-site clusters indicate a net electron attraction leading to the formation of bound electron pairs and possible condensation at finite temperature for $U < U_c(T)$ [2,3]. This pairing mechanism in the 4-site cluster, at $1/8$ (optimal) hole doping ($\langle N \rangle \approx 3$) away from half filling, exists when the onsite Coulomb interaction U is less than an analytically obtained critical value, $U_c(T = 0) = 4.584$ (in units of the hopping parameter t). This critical value, first reported in Ref. [3], is temperature dependent and can be associated with an energy gap (order parameter) which becomes negative below $U_c(T)$ implying that it is more energetically favorable to have a bound pair of electrons (or holes) compared to two unpaired ones at an optimal chemical potential (or doping level) $\mu = \mu_P = 0.658$. Above this critical value $U_c(T)$, there is a Mott–Hubbard like gap that exists when the average particle number $\langle N \rangle \approx 3$; this gap decreases monotonically as U decreases and vanishes at $U_c(T)$. The vanishing of the gap indicates the *onset of pair formation*. There is an interval (width) around μ_P, where the pairing phase competes with a phase (having a high magnetic susceptibility) that suppresses pairing at 'moderate' temperatures.

An enlarged view of the T-μ phase diagram, for the planar 4-site cluster near μ_P, is shown in Figure 10.1. This exact phase diagram (at $U = 4$) in the vicinity of the optimally doped ($N \approx 3$) regime has been constructed using the ideas described in the text and Ref. [2]. The electron pairing temperature, T_c^P, identifies the onset of charge pairing. As temperature is further lowered, spin pairs begin to form at T_s^P. At this temperature (with zero magnetic field), spin susceptibilities become very weak indicating the disappearance of the $\langle N \rangle \approx 3$ states. Below this spin pairing temperature T_s^P, only paired states are observed to exist having a certain rigidity, so that a nonzero magnetic field or a finite temperature is required to break the pairs. From a detailed analysis, it becomes evident that the system is on the verge of an instability; the paired phase competing with a phase that suppresses pairing which has a high, zero-field magnetic susceptibility. As the temperature is lowered, the number of $\langle N \rangle \approx 3$ (unpaired) clusters begins to decrease while a mixture of (paired) $\langle N \rangle \approx 2$ and $\langle N \rangle \approx 4$ clusters appears. Interestingly, the

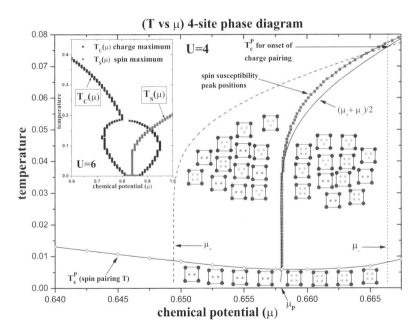

Figure 10.1. The T-μ phase diagram near $\mu_P = 0.658$ ($\langle N \rangle \approx 3$) at $U = 4$ for the 4-site cluster. The inset shows a corresponding section (at a different scale) of the T-μ phase diagram for $U = 6$. For $U = 4$, note how the paired states condense at low temperature ($T_s^P \leqslant 100$ K if $t = 1$ eV) with a nonzero pair binding energy, while at higher temperatures, unpaired states begin to appear. This picture supports the idea that there is inhomogeneous, electronic phase separation here. When U is higher than $U_c(0) = 4.584$ (see inset for $U = 6$), these inhomogeneities disappear and a Mott–Hubbard like stable, paramagnetic, insulating region results around optimal doping. Note that there is charge-spin separation at $U = 6$ in this gapped, insulating region (charge and spin susceptibility peaks are denoted by $T_c(\mu)$ and $T_s(\mu)$), as discussed in Ref. [2].

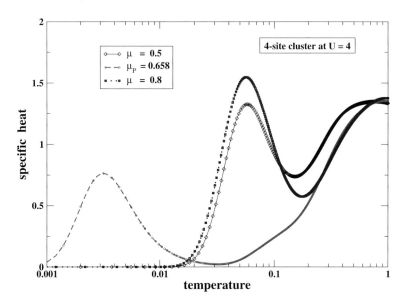

Figure 10.2.   Specific heat vs. temperature at $U = 4$, calculated in the grand canonical ensemble for the 4-site cluster at several doping values near the critical doping, $\mu_P \approx 0.658$. Note how the low temperature peak shifts to higher temperatures when the doping is changed from its critical value.

critical doping μ_P (which corresponds to a filling factor of $1/8$ hole-doping away from half filling), where the above pairing fluctuations take place when $U < U_c(T)$, is close to the doping level near which numerous intriguing properties have been observed in the hole-doped HTSCs.

Specific heat calculations (Figure 10.2), associated with energy fluctuations, also provide further support for an electronic phase change at low temperature. As seen in this figure, there is a well separated, low temperature peak at $\mu_P = 0.658$ (around 40 K, if the hopping parameter is set to 1 eV and U to 4 eV). This peak, which shifts to higher temperatures when the doping level is different from critical doping, is due to fluctuations between paired states ($\langle N \rangle = 2$ and $\langle N \rangle = 4$). This low temperature peak is in agreement with specific heat experiments carried out for the HTSCs [4], and is a manifestation of the near degeneracy of the states in the neighborhood of critical doping μ_P and onset of condensation.

10.4.   T-U phase diagram and pressure effects

The T-U phase diagram in Figure 10.3 illustrates the effects of the Coulomb repulsion U on possible binding of electron charge and spin in the 4-site cluster. The charge and spin pairing phase boundaries, $T_c^P(U)$ and $T_s^P(U)$, shown in Figure 10.3 are constructed by monitoring the vanishing of the charge and the spin gaps [2] near optimal doping as functions of T and U. For temperatures $T_s^P(U) \leqslant T \leqslant T_c^P(U)$, there is bound charge $2e$ and decoupled spins. Below the lower curve, when $T \leqslant T_s^P(U)$, the spin degrees are also bounded and a finite applied magnetic field is needed to break them [14,2].

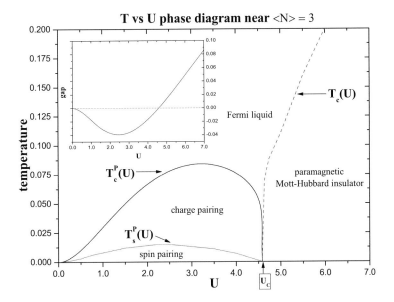

Figure 10.3. *T* vs. *U* phase diagram for the optimally doped 4-site clusters, based on our exact calculations. $T_c^P(U)$ denotes the temperature at which the charge gap (see text) vanishes. Note that the pressure effects can be related to the behavior of $T_s^P(U)$, below which the spins are paired. Increasing the pressure is equivalent to decreasing U, while t is held constant at $t = 1$. The inset shows the charge gap as a function of U at zero temperature. A negative charge gap implies charge pairing. Note that when $U > U_c$, there is neither charge nor spin pairing. Instead, a MH like insulating phase, with a positive charge gap at low temperature, is found near $\langle N \rangle \approx 3$; its phase boundary ($T_c(U)$) is denoted by the dashed line.

In Figure 10.3 for $U > U_c$, notice the existence of electron–hole pairing in a Mott–Hubbard like insulating region (below temperature $T_c(U)$ denoted by the dashed line) away from half filling [19]. The inset in Figure 10.3 shows the variation of the charge gap, $E(2) + E(4) - 2E(3)$, as a function of U where $E(N)$ refers to the canonical energy for N electrons at $T = 0$. When this gap is negative, pairing is favored as discussed in Ref. [2].

Figure 10.3 can also be used to understand pressure related effects, noting that increasing the pressure is equivalent to decreasing U/t. Hence, the increase of $T_s^P(U)$ with decreasing U, while holding t constant at 1 (in Figure 10.3) when $2.5 \leqslant U \leqslant U_c$, reproduces the superconducting transition temperature (STT) vs. pressure p behavior in organic superconductors [22] and optimally and nearly optimally doped HTSC materials [29], indicating a significant role of pair binding in enhancing the STT. Notice that, at small enough U, $T_s^P(U)$ decreases with decreasing U (i.e., under increasing pressure) as shown in Figure 10.3. This might explain why the pressure strongly depresses the STT across some families of alkali doped, fullerene as well as organic superconductors [20].

10.5. Unpaired, dormant magnetic state

Another intriguing fact emerging from the exact thermal studies of the 4-site clusters is the existence of a dormant magnetic state (unpaired states with $\langle N \rangle = 3$) with a high magnetic

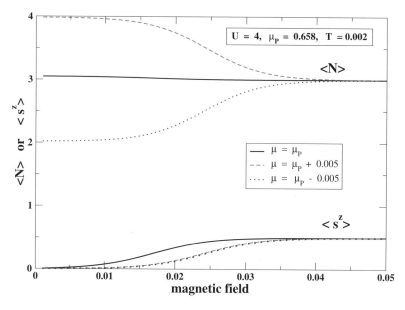

Figure 10.4.   Variation of electron number $\langle N \rangle$ and magnetization $\langle s^z \rangle$ as a function of external magnetic field for several values close to critical doping $\mu_P = 0.658$ at $T = 0.002$ and $U = 4$ for the 4-site cluster. Note how the $\langle N \rangle = 3$ clusters get stabilized in a nonzero magnetic field at low temperature. These results are reminiscent the recent observation of a magnetic state near optimal doping in hole-doped La cuprates [30].

susceptibility. At rather low temperature $T \leqslant T_s^P$, this state is dormant. However, a small magnetic field or a change in chemical potential can stabilize it over the paired states $\langle N \rangle \approx 2, 4$ as seen from Figure 10.4 and the calculated grand canonical probabilities (not shown). The magnetization curves indicate the fact that the $\langle N \rangle \approx 3$, unpaired states are easier to magnetize with an infinitesimal field. The variation of the magnetic field mimics the doping to some extent here. Small changes in doping (at zero field) can also switch the system from one state to another with a different $\langle N \rangle$. These may be compared to some recent experimental results reported in Ref. [30], where a magnetic (and non-superconducting) state has been observed near 1/8 hole-doping in $La_{2-x}Sr_xCuO_{4+y}$. This system is said to be on the verge of an instability, surprisingly similar to the phase space region at $\langle N \rangle \approx 3$ (i.e., near optimal doping away from half-filling – see Figures 10.1 and 10.4) at low temperature. The physics behind this can be directly tied to the correlation driven instability near optimal doping when $U \leqslant U_c(T)$.

10.6.   Linked 4-site clusters

In order to monitor size effects, it is important to carry out numerical calculations for clusters with different topologies and sizes [13,23]. Figure 10.5 illustrates one such set of calculations of charge gaps carried out for a 8-site cluster (2×4 ladder), where the hopping term or coupling c between the two squares was allowed to be different from the coupling within a given square. The pairing fluctuations, that are seen for the 4-site cluster, exist even for these ladders near half-filling, at optimal doping ($\langle N \rangle \approx 7$), and

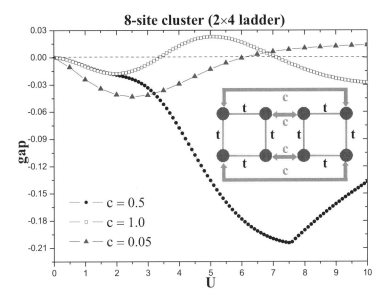

Figure 10.5.   Charge gaps for the 2×4 cluster at $T = 0$ for various couplings **c** between the squares. The doping level is one electron off half-filling and the couplings t within the squares are set to 1, as indicated. There is an effective electron–electron attraction in the negative charge gap regions.

most of the trends observed for the 4-site clusters, such as the MH like charge gaps and vanishing of such gaps at critical U values, remain valid here. The fluctuations that occur here at optimal doping are among the states with $\langle N \rangle \approx 6$, 7 and 8 electrons. Clearly, the dormant magnetic state corresponds to $\langle N \rangle \approx 7$. Thermal and quantum fluctuations in the density of holes between the clusters (for $U < U_c(0)$) make it energetically more favorable to form pairs. In this case, snapshots of the system at relatively low temperatures and at a critical doping level (such as μ_P in Figure 10.1) would reveal phase separation and equal probabilities of finding hole-rich or hole-poor clusters in the ensemble.

10.7.   Summary

In summary, these exact Hubbard cluster calculations, carried out at various fillings, show the existence of charge and spin pairing, electronic phase separation, pseudogaps, and condensation driven by electron correlations. These small clusters demonstrate a rich variety of properties which can be tuned by electron or hole doping. Such exact cluster studies may be useful in understanding the phase diagrams and superconductivity driven by purely electronic means (such as in CNs). Furthermore, it is quite surprising to see the number of properties that these exact clusters share with the HTSCs. This may be, at least in part, due to the fact that in all these 'bad' metallic high T_c materials, short-range correlations play a key role. Other exact cluster studies, carried out using the method outlined here, are beginning to yield important insights into numerous problems such as Nagaoka type magnetic instabilities [32] and frustrated nanomagnets. Some of these insights may be useful in designing ultra-small and true nanoscale devices which could compete with those based on multilayers in the future.

Acknowledgements

This work was done in collaboration with my colleagues, Armen Kocharian, Jim Davenport, Kalum Palandage, Tun Wang.

References

[1] G.W. Fernando, A.N. Kocharian, K. Palandage, Tun Wang, J.W. Davenport, Phys. Rev. B **75**, 085109 (2007).
[2] A.N. Kocharian, G.W. Fernando, K. Palandage, J.W. Davenport, Phys. Rev. B **74**, 024511 (2006).
[3] A.N. Kocharian, G.W. Fernando, K. Palandage, J.W. Davenport, J. Magn. Magn. Mater. **300**, e585 (2006).
[4] G.V.M. Williams, J.L. Tallon, J.W. Loram, Phys. Rev. B **58**, 15053 (1998).
[5] V.J. Emery, S.A. Kivelson, O. Zachar, Phys. Rev. B **56**, 6120 (1997).
[6] Y. Zha, V. Barzykin, D. Pines, Phys. Rev. B **54**, 7561 (1996).
[7] E. Dagotto, New Jour. Phys. **7**, 67 (2005).
[8] K.K. Gones, A.N. Pasupathy, A. Pushp, Y. Ando, A. Yazdani, Nature **447**, 569 (2007).
[9] H. Shiba, P.A. Pincus, Phys. Rev. B **5**, 1966 (1972).
[10] J. Callaway, D.P. Chen, R. Tang, Phys. Rev. B **35**, 3705 (1987).
[11] Claudius Gros, Phys. Rev. B **53**, 6865 (1996).
[12] F. Lopez-Urias, G.M. Pastor, Phys. Rev. B **59**, 5223 (1999).
[13] E. Dagotto, Rev. Mod. Phys. B **66**, 763 (1994).
[14] A.N. Kocharian, Joel H. Sebold, Phys. Rev. B **53**, 12804 (1996).
[15] R. Schumann, Ann. Phys. **11**, 49 (2002).
[16] J.E. Hirsch, Phys. Rev. B **67**, 035103 (2003).
[17] J. Bonca, P. Prelovsĕk, Int. J. Mod. Phys. B **17**, 3377 (2003).
[18] M. Cini, Topics and Methods in Condensed Matter (Springer-Verlag, Berlin, 2007).
[19] A.N. Kocharian, G.W. Fernando, K. Palandage, J.W. Davenport, 2007; cond-mat/0701022.
[20] T. Yildirim, O. Zhou, J.E. Fischer, N. Bykovetz, R.A. Strongin, M.A. Cichy, A.B. Smith III, C.L. Lin, R. Jelinek, Nature **360**, 568 (1992).
[21] A. Kasumov, et al., Phys. Rev. B **68**, 214521 (2003).
[22] M.A. Tanatar, T. Ishiguro, S. Kagoshima, N.D. Kushch, E.B. Yagubskii, Phys. Rev. B **65**, 064516 (2002).
[23] S. Belluci, M. Cini, P. Onorato, E. Perfetto, J. Phys.: Condens. Matter **18**, S2115 (2006);
W.-F. Tsai, S.A. Kivelson, Phys. Rev. B **73**, 214510 (2006);
S.R. White, S. Chakravarty, M.P. Gelfand, S.A. Kivelson, Phys. Rev. B **45**, 5062 (1992);
R.M. Fye, M.J. Martins, R.T. Scalettar, Phys. Rev. B **42**, R6809 (1990);
N.E. Bickers, D.J. Scalapino, R.T. Scalettar, Int. J. Mod. Phys. B **1**, 687 (1987).

[24] J.M. Tranquada, B.J. Sternlieb, J.D. Axe, Y. Nakamura, S. Uchida, Nature **375**, 561 (1995).

[25] T. Koma, H. Tasaki, Phys. Rev. Lett. **68**, 3248 (1992).

[26] P.W. Anderson, Science **235**, 1196 (1987).

[27] X. Xu, S. Yin, R. Moro, W.A. de Heer, Phys. Rev. Lett. **95**, 237209 (2005).

[28] I. Baldea, H. Koppel, L.S. Cederbaum, Eur. Phys. J. B **20**, 289 (2001).

[29] Xiao-Jia Chen, Viktor V. Struzhkin, Russell J. Hemley, Ho-kwang Mao, Chris Kendziora, Phys. Rev. B **70**, 214502 (2004).

[30] H.E. Mohottala, B.O. Wells, J.I. Budnick, W.A. Hines, C. Niedermayer, L. Udby, C. Bernard, A.R. Moodenbaugh, Fang-Cheng Chou, Nature Materials **5**, 377 (2006).

[31] Y. Itoh, M. Matsumara, H. Yamagata, J. Phys. Soc. Jpn. **66**, 3383 (1997).

[32] Y. Nagaoka, Sol. State Comm. **3**, 409 (1965);
Y. Nagaoka, Phys. Rev. **147**, 392 (1965).

Subject Index